T0306016

Innovative Thermoelectric Materials

Polymer, Nanostructure and Composite Thermoelectrics

Innovative Thermoelectric Materials

Polymer, Nanostructure and Composite Thermoelectrics

Editors

Howard E Katz
Theodore O Poehler

Johns Hopkins University, USA

Imperial College Press

Published by

Imperial College Press
57 Shelton Street
Covent Garden
London WC2H 9HE

Distributed by

World Scientific Publishing Co. Pte. Ltd.
5 Toh Tuck Link, Singapore 596224
USA office: 27 Warren Street, Suite 401-402, Hackensack, NJ 07601
UK office: 57 Shelton Street, Covent Garden, London WC2H 9HE

British Library Cataloguing-in-Publication Data
A catalogue record for this book is available from the British Library.

ISBN 978-1-78326-605-0

In-house Editors: R. Raghavarshini/Mary Simpson

Typeset by Stallion Press
Email: enquiries@stallionpress.com

Printed in Singapore

Contents

List of Contributors

David R. Brown
Materials Science
California Institute of Technology
1200 East California Boulevard
Pasadena
California
91125
USA

M. Chabinyc
Materials Department
University of California
Santa Barbara
California
93106-5050
USA

Gang Chen
Department of Mechanical Engineering
Massachusetts Institute of Technology
77 Massachusetts Avenue
Cambridge, MA
20139
USA

Nelson E. Coates
Molecular Foundry
Lawrence Berkeley National Laboratory
Berkeley, CA
94720
USA

Anne Glaudell
Materials Department
University of California
Santa Barbara, CA
93106-5050
USA

Howard E. Katz
Department of Materials Science and Engineering
Johns Hopkins University
3400 North Charles Street
Baltimore, MD
21228
USA

Robert M. Ireland
Department of Materials Science and Engineering
Johns Hopkins University
3400 North Charles Street
Baltimore, MD
21228
USA

Gun-Ho Kim
Department of Mechanical Engineering
University of Michigan
Ann Arbor, MI
48109-2125
USA

Sangyeop Lee
Department of Mechanical Engineering
Massachusetts Institute of Technology
77 Massachusetts Avenue
Cambridge, MA
20139
USA

Kevin P. Pipe
Department of Mechanical Engineering
University of Michigan
Ann Arbor, MI
48109-2125
USA

Theodore O. Poehler
Department of Materials Science and Engineering
Johns Hopkins University
3400 North Charles Street
Baltimore, MD
21228
USA

Ruth Schlitz
Materials Department
University of California
Santa Barbara, CA
93106-5050
USA

G. Jeffrey Snyder
Materials Science
California Institute of Technology
1200 East California Boulevard
Pasadena, CA
91125
USA

Jeffrey J. Urban
Molecular Foundary
Lawrence Berkeley National Laboratory
Molecular Foundary
Lawrence Berkeley National Laboratory
Berkeley, CA
94720
USA

D. Gregory Walker
Department of Mechanical Engineering
Vanderbilt University
2301 Vanderbilt Place
Nashville, TN
37235
USA

Preface*

Organic conductors, polymers, and composites have clearly demonstrated considerable potential as electronic and photonic materials for a variety of applications that previously depended on the use of inorganic compounds. The progress on conducting organic materials and polymers beginning with the discovery of polyacetylene has resulted in a host of devices being demonstrated including organic and polymer-based field effect transistors, organic solar cells, sensors, batteries, and many passive components. A logical progression of this evolution is for these materials to be pursued as thermoelectric materials for conversion of thermal energy into useful electric power predominantly in moderate temperature range. A combination of organic, polymeric, and nanostuctured materials have created a rich array of materials that can be combined to form thermoelectric systems that have begun to display operating characteristics that may lead practical alternatives to current inorganic thermoelectric alloy systems for appropriate applications.

The volume begins with a broad perspective of thermoelectric materials by Poehler and Katz extending from the early work on bulk intermetallic compounds such as the widely investigated compound Bi_2Te_3 and Bi_2Te_3 alloys capable of higher thermoelectric power generation because of reduced thermal conductivity due to greater

*Howard Katz is grateful to the Department of Energy, Office of Basic Energy Sciences, Grant Number DE-FG02-07ER46465, for supporting his contributions.

acoustic phonon scattering in the alloys. A summary of solid-state chemistry approaches to synthesize better thermoelectric materials through structural and chemical methods in ternary and quaternary systems are described. A series of alloys are illustrated including half-Heusler alloys, skutterides, and chalcogenide compounds that have proven to be thermoelectrics with low thermal conductivity and high power factors. New materials are described using a series of structural innovations comprising nanostructures, composites, and polymer materials that succeed in modifying dimensionality, decoupling electrical and thermal conductivity, and extending the field of thermoelectric materials beyond the traditional inorganic compounds. The chapter concludes with the recognition that there is much work to accomplish to fulfill the promise of these new materials and develop an understanding of theory to extend the effectiveness of these new materials.

The second chapter in the volume by Lee and Chen discusses how the introduction of nanostructures in thermoelectric materials has led to significant improvements in the thermoelectric figure of merit by modifying transport properties such as carrier concentration and tailoring band structure yielding a density of states that selectively promotes the transport of energetic carriers as well as carrier scattering. It describes some of the advantages and outcomes in polymer-based thermoelectric materials, as well as in polymer–inorganic nanomaterials that hold promise for enhancement of the electronic properties by introduction of nanostructures to tailor the electron and phonon transport more effectively than prior methods. The advantages of nanostructures are a major focal point of the discussion, but a number of other tactics are also described that have potential for improving the performance of bulk thermoelectric materials as well.

In the next chapter by Schlitz, Glaudell, and Chabinyc the authors concentrate on the thermoelectric performance and transport models employed to comprehend organic polymers and solution processed organic semiconductors. Thermolectric behavior in materials synthesized by vapor deposition techniques has recently been the subject of a study by Walzer *et al.* particularly with respect to injection layers in organic light emitting diodes (OLEDs). *Chem*

Rev. **107**, 1233–1271 (2007). This work provides a comprehensive summary of recent progress in n- and p-type materials together with a thoughtful assessment of barriers to be overcome to enhance organic thermoelectric materials.

In the fourth chapter by Urban and Coates the focal point of the discussion is the design rules for polymer-based thermoelectric nanocomposites. A universal aspect of improvements in thermoelectric materials performance in recent years is the greater complexity used in their production, and the most recent manifestation of this is present in polymer nanocomposite materials. In organic–inorganic nanoscale composites carrier filtering or energy filtering by selection of carriers with high energies, or power factor enhancement through trap states can produce electronic carriers improved thermoelectric performance. The authors have explained the primary rules for creating the optimum thermoelectric transport in polymer composite materials that have the potential to lead to design unique materials of vastly improved performance.

The next chapter by Kim and Pipe reviews the role of dopant atoms or molecules in establishing the optimum thermoelectric parameters so as to maximize the thermoelectric figure of merit ZT. In thermoelectric materials, the opposite dependence of the Seebeck coefficient and the electrical conductivity on carrier concentration implies that there is an optimum concentration that will maximize the power factor $S^2\sigma$. In order to achieve the optimum carrier concentration, the density of impurity or dopant atoms must adjusted, and may affect the electronic carrier mobility. In contrast to inorganic semiconductors, where mobility declines at increasing carrier concentration through higher impurity scattering, in organic semiconductors the mobility increases with carrier concentration until the dopant volume becomes quite large. The authors describe a model that demonstrates the importance of minimizing dopant volume in maximizing ZT, and simultaneously enhancing the values of the three parameters that constitute ZT.

The sixth chapter by Ireland and Katz describes the potential for improved thermoelectric performance from complex polymer–inorganic composites. Polymer–inorganic composites have been prepared by techniques including physical mixing, solution processing, and *in situ* synthesis of polymers and inorganic materials with improved properties compared to separate components. Integrating conducting organic compounds in conjugated polymers is anticipated to result in the composite material having improved electrical conductivity and a higher Seebeck coefficient with the polymer matrix containing the inorganic compound exhibiting a low thermal conductivity. Highly ordered composites are expected to have enhanced thermoelectric performance as a result of highly engineered semiconductors incorporating better current pathways, morphological confinement, energy filtering, and alignment and interfaces with selective scattering.

The next chapter by Brown and Snyder describes the role that structural entropy can play in strongly enhancing the thermopower of complex thermoelectric materials yielding a new approach to creating systems with high ZT and wide applications for these thermoelectric materials. In particular, the authors have demonstrated that coupling of a continuous phase transition to carrier transport in CuSe results in a striking peak in thermopower, increased scattering of electrons and phonons, and a doubling of ZT. The correlation of improved performance in the material with the nature of the phase transition is indicative of the entropy being associated with ion ordering. This implies that a new mechanism for high thermoelectric performance may be recognized and utilized as an approach to enhancing function.

The chapter by Walker describes some of the predominant techniques for modeling and predicting the transport properties of contemporary thermoelectric materials. From these models, we enhance our ability to comprehend the physical characteristics of materials that influence when a material will function effectively as a thermoelectric material. The modeling involves both the electronic and thermal systems with a concentration on noncontinuum models that include both particle-oriented techniques such as the Boltzmann model, quantum-based techniques, for instance, the nonequilibrium Green's function method, and atomistic simulation. Notwithstanding that the energy-bearing carriers, electrons and phonons, are clearly joined, the two systems are treated independently inasmuch as the coupling is generally minimal. Methods such as the nonequilibrium Green's function model appear to match experimental data well when predicting thermoelectric parameters like the Seebeck coefficient.

Chapter 1

Innovative Thermoelectric Materials

Theodore O. Poehler and Howard E. Katz

1.1 Thermoelectric Effect Mechanisms

1.1.1 Introduction

In recent years, there has been an increasing emphasis on meeting the demand for electrical power by using renewable noncarbon-based sources. The conventional alternative sources that are identified as candidates for these applications are solar, geothermal, and wind energy. Notwithstanding the attractiveness of these environmentally friendly sources, they are currently providing only a very modest amount of energy in developed countries while having a somewhat limited attractiveness because of their high cost and the uncertainty in being able to generate the quantity of electrical power that is required to make an impact on the energy utilization of major industrial economies.

A less recognized source of available electrical energy is the immense amount of waste heat that currently is not employed despite its great potential and prospectively low cost. Approximately two thirds of the energy used in generating electric power by conventional means is converted to heat and ultimately dissipated. In some combined electrical and heat generation systems, about one half of the energy is usefully converted, but as power plants are increasingly becoming more remote from areas where the heat may be directly utilized that fraction diminishes. A promising methodology

1

for converting waste heat to useful electrical energy is thermoelectric energy generation technology.

Thermoelectric energy generation directly converts heat into electrical power using the nonuniform distribution of heat in a medium to establish an electrical voltage[1,2] that may be used as a power supply for various applications ranging from thermoelectric coolers to climate control systems.[2,3] Thermoelectric power is of interest not only in small-scale applications as mentioned above, but also in addition may be used to deliver meaningful amounts of electrical power to widespread locales and communities in developing countries that do not have the infrastructure to distribute centrally generated electrical power throughout a large region. Somewhat analogous to the use of distributed communications systems based on satellite transmissions in regions without an expensive communications structure, the local electrical power might be developed locally by thermoelectric systems in territory far removed from population centers where electric energy might make a significant improvement in the nature of life for local residents.

Thermoelectric power sources have been familiar for many years, but have not been viewed as an important solution to energy problems because of their usual limited capacity and low efficiency. However, it is possible that a new vision for the application of thermoelectric power sources would be feasible if the thermoelectric technology advances in the same manner as has been occurring with improved solar energy sources.

1.1.2 Fundamental characteristics of thermoelectric effects

The basic source of the thermoelectric effect is the transport of electrical charge and energy in conducting materials.[4] The equilibrium flow of electric charge and thermal energy establishes a potential difference between the hot and cold ends of the material. In more specific terms, the temperature gradient induces a change in the equilibrium distribution function $f(\mathbf{k})$ that describes the distribution of electrons among various allowed energy levels for different magnitudes of the wave vector $\mathbf{k}$. The change in the distribution function to a new value may cause motion of both electrical charges and heat.

1.1.2.1 *Electrostatic equilibrium*

In an equilibrium semiconductor with no applied electric field, but with a thermal gradient, the distribution of electrons and holes among the available states is different on the hot side from that on the cold side, exclusively from the spatial variation in the Fermi distribution function for the different temperatures of the densities of states at the opposite ends. This creates a density gradient in the number of carriers in each state at the opposing regions. Therefore, carriers diffuse in response to this gradient, effectively raising the carrier temperature on the cold side also. However, the carriers on the cold side reach an elevated electrochemical potential because of the accumulating voltage from the transfer of the carriers. Electrostatic equilibrium is reached when the sum of the free energies of all the carriers, including contributions from entropy increases from populating higher energy states on the cold side and the effect of the voltage that also builds up between the hot and cold sides, reaches a minimum. This voltage may be expressed as the area charge density per specific capacitance at each position of the sample. It is assumed that this movement of charge does not actually change the phonon temperature of either side; external thermal power maintains the temperature difference independently of the movement of the carriers.

1.1.2.2 *Zero net transport*

Diffusion of the hot carriers occurs at some rate that depends on the activation energy (if any) of carriers moving from site-to-site, the probabilities that the first and second site would be occupied according to the distribution function, and the scattering lifetime. The concentration gradient of hot carriers between the hot and cold side drives diffusion of the carriers from hot to cold sides. The higher the diffusion rate, the higher the current. The net current will be zero when a voltage builds up that would drive an equal and opposite drift current, according to Ohm's law. These opposite currents would "flow" indefinitely, because even though the system is static electrically, it is not static thermally. There is a continual resupplying of heat to the hot side to generate new hot carriers. The Seebeck coefficient is a parameter that quantifies the voltage per unit temperature generated in a dynamic system driven by the heat flow due to a thermal gradient that is, in turn, balanced in equilibrium from the electric field that opposes the charge flow due to the thermal gradient.

1.1.3 Boltzmann equation

In an equilibrium system, no transport of charge or energy occurs in the distribution function if $f(\mathbf{k})$ is symmetric in $\mathbf{k}$ space given that there are equivalent amounts of carriers with momentum that are equal but opposite so the current is zero. In order that there will be transport of either charge or energy the distribution function must be asymmetric in $\mathbf{k}$ space. We can describe the distribution function $f(\mathbf{k}, \mathbf{r}, t)$ as the probability at a time t that a state having a wave vector $\mathbf{k}$ will be occupied at a point in the material designated by the position vector $\mathbf{r}$. If there is a force $\mathbf{F}$ on the electrons where $\mathbf{F} = \hbar d\mathbf{k}/dt$, then the total rate of change of the distribution function will be

$$\frac{df}{dt} = \frac{\partial f}{\partial t} + \hbar^{-1}\mathbf{F} \cdot \nabla_k f + \mathbf{v} \cdot \nabla f. \qquad (1.1)$$

An electron subject to a force $\mathbf{F}$ will change position from $\mathbf{r}$ and a wave vector $\mathbf{k}$ to $\mathbf{r} + \mathbf{v}\, dt$ and a wave vector $\mathbf{k} + \hbar^{-1}\mathbf{F}\, dt$ at a time $t + dt$ where $\mathbf{v}$ is the velocity vector.

The total rate of change in f is caused by collisions so the rate of change resulting from collisions is given by $df/dt|_{coll}$ thus we obtain the following equation:

$$\frac{\partial f}{\partial t} = \frac{df}{dt}\Big]_{\mathrm{coll}} - \hbar^{-1}\mathbf{F} \cdot \nabla_k f - \mathbf{v} \cdot \nabla f. \qquad (1.2)$$

This is Boltzmann's equation that is the primary relation used as the basis for describing all transport phenomena.[5]

In the time-independent case we have $\frac{\partial f}{\partial t}$ equal to zero, and

$$\frac{df}{dt}]_{\text{coll}} = \hbar^{-1}\mathbf{F}\cdot\nabla_k f - \mathbf{v}\cdot\nabla f. \tag{1.3}$$

If there are no external fields or temperature gradients then the distribution will reach thermal equilibrium through collisions in a certain time assuming that the distribution is not already in equilibrium. Generally, the relaxation by collisions occurs at a rate proportional to the derivative of the distribution function. The time to reach equilibrium is termed the relaxation time, and is given by the following equation:

$$\frac{df}{dt}]_{\text{coll}} = -\frac{f - f_o}{\tau}, \tag{1.4}$$

where f_o is the equilibrium distribution function.

With this form for the collision time, Boltzmann's equation is

$$f = f_o - \tau\hbar^{-1}\mathbf{F}\cdot\nabla_k f - \tau\mathbf{v}\cdot\nabla f. \tag{1.5}$$

If we assume that the difference between f and f_o is small, then f can be replaced by f_o in Eq. (1.4) so the Boltzmann equation can be expressed as

$$f = f_o - \tau\hbar^{-1}\mathbf{F}\cdot\nabla_k f_o - \tau\mathbf{v}\cdot\nabla f_o. \tag{1.6}$$

Inasmuch as f_o is a function of energy ε and $\nabla_k\varepsilon = \hbar\mathbf{v}$ the Boltzmann equation may be written as

$$f = f_o - \tau\frac{\partial f_o}{\partial\varepsilon}(\mathbf{F}\cdot\mathbf{v}) - \tau\mathbf{v}\cdot\nabla f_o. \tag{1.7}$$

If we have a uniform electric field $\mathbf{E}$ directed along a single axis, x, e is the electronic charge, and no temperature gradient dT/dx along the same axis then the distribution function f becomes

$$f = f_o + \tau e \mathbf{E} v_x \frac{\partial f_o}{\partial\varepsilon}. \tag{1.8}$$

1.1.3.1 *Current density*

In a system with the distribution function of Eq. (1.8) the current density associated with these carriers is given by

$$\mathbf{J} = -e \int N(\mathbf{k}) \mathbf{v} f(\varepsilon) d\mathbf{k}, \tag{1.9}$$

$$\mathbf{J} = -e \int N(\varepsilon) \mathbf{v} f(\varepsilon) d\varepsilon, \tag{1.10}$$

where $N(\mathbf{k})$ and $N(\varepsilon)$ are the densities of states expressed as either a function of k or ε, respectively.

By inserting the distribution function (1.8) along the x-axis in the expression for the current density and setting $N(\mathbf{k})$ equal to $V/8\pi^3$, where V is the volume of the crystal, we find

$$J_x = -e \int N(\mathbf{k}) \mathbf{v} f(\varepsilon) d\mathbf{k}' = -\frac{e^2\,\mathbf{E}}{4\pi^3} \int \tau v_x^2 \frac{\partial f_o}{\partial \varepsilon} d\mathbf{k}'. \tag{1.11}$$

1.1.3.2 *Thermal conduction*

If we consider the effect of both a constant electric field along the x-axis and a temperature gradient in the same direction, then the modified distribution function in Eq. (1.3) becomes

$$f = f_o + \tau e \mathbf{E} v_x \frac{\partial f_o}{\partial \varepsilon} - \tau v_x \frac{\partial f_o}{\partial x}. \tag{1.12}$$

If no external fields are applied to the material then energy is transported in the direction of the thermal gradient as a result of an inhomogeneous distribution of the electronic carriers of different energies and is responsible for thermal conduction. If there are thermal gradients it is necessary to utilize Boltzmann's equation to determine the relevant equations, which further determine the relations governing the electrical current density and the thermal energy flow.

We will express

$$\frac{\partial f_o}{\partial x} = \frac{\partial f_o}{\partial T} \frac{\partial T}{\partial x}, \tag{1.13}$$

where the Fermi energy ε_F is a function of T, so

$$\frac{\partial f_o}{\partial T} = kT \frac{\partial f_o}{\partial \varepsilon} \frac{d}{dT} \left(\frac{\varepsilon - \varepsilon_F}{kT} \right), \qquad (1.14)$$

$$\frac{\partial f_o}{\partial T} = - \left\{ T \frac{d}{dT} \left(\frac{\varepsilon_F}{T} \right) + \frac{\varepsilon}{T} \right\} \frac{\partial f_o}{\partial \varepsilon}. \qquad (1.15)$$

Thus we represent the distribution function in Eq. (1.12) as

$$f = f_o + \tau v_x \frac{\partial f_o}{\partial \varepsilon} \left[e\mathbf{E}_x + \frac{dT}{dx} \left\{ T \frac{d}{dT} \left(\frac{\varepsilon_F}{T} \right) + \frac{\varepsilon}{T} \right\} \right]. \qquad (1.16)$$

Using this distribution function we may calculate the electrical current density in the x direction under the conditions of a thermal gradient as well as an electric field which yields

$$J_x = -\frac{e}{4\pi^3} \int \tau v_x^2 \frac{\partial f_o}{\partial \varepsilon} \left[e\mathbf{E}_x + \left\{ T \frac{d}{dT} \left(\frac{\varepsilon_F}{T} \right) + \frac{\varepsilon}{T} \right\} \frac{dT}{dx} \right] d\mathbf{k}'. \qquad (1.17)$$

We may also calculate the thermal current W_x resulting from the energy transport along the same axis as

$$W_x = -\frac{1}{4\pi^3} \int \tau v_x^2 \frac{\partial f_o}{\partial \varepsilon} \left[e\mathbf{E}_x \varepsilon + \left\{ \varepsilon T \frac{d}{dT} \left(\frac{\varepsilon_F}{T} \right) + \frac{\varepsilon^2}{T} \right\} \frac{dT}{dx} \right] d\mathbf{k}'. \qquad (1.18)$$

If we have a plain band system with a single band represented by a simple effective mass m_e, the expressions in Eqs. (1.17) and (1.18) can be further reduced to the following relations:

$$J_x = \frac{ne}{m_e} \left[\left\{ e\mathbf{E}_x + T \frac{d}{dT} \left(\frac{\varepsilon_F}{T} \right) \frac{dT}{dx} \right\} \langle \tau \rangle + \frac{1}{T} \frac{d\mathbf{T}}{dx} \langle \varepsilon \tau \rangle \right], \qquad (1.19)$$

$$W_x = -\frac{n}{m_e} \left[\left\{ e\mathbf{E}_x + T \frac{d}{dT} \tau \left(\frac{\varepsilon_F}{T} \right) \frac{dT}{dx} \right\} \langle \tau \varepsilon \rangle + \frac{1}{T} \frac{d\mathbf{T}}{dx} \langle \varepsilon^2 \tau \rangle \right]. \qquad (1.20)$$

In these equations, the values for the scattering time, $\langle \tau \rangle$, as well as the products $\langle \tau \varepsilon \rangle$ and $\langle \tau \varepsilon^2 \rangle$ are expressed as suitable average quantities depending on the type of materials where for nondegenerate semiconductors an appropriate weighting function is used, i.e.,

$\tau = a\varepsilon^{-s}$. For degenerate semiconductors or metals, the average values are given by $\langle\tau\rangle = \tau_F$, $\langle\tau\varepsilon\rangle = \tau_F\varepsilon_F$ and $\langle\tau\varepsilon^2\rangle = \tau_F\varepsilon_F^2$.

1.1.4 Thermoelectric effect — Seebeck coefficient

If we consider a case when no current flows, that is where Eq. (19) is set to zero, then we will have an electric field established in the material as a result of the thermal gradient.

Equation (1.19) will become

$$E_x = -\frac{1}{\langle\tau\rangle e}\left\{T\frac{d}{dT}\left(\frac{\varepsilon_F}{T}\right)\langle\tau\rangle + \frac{1}{T}\langle\varepsilon\tau\rangle\right\}\frac{dT}{dx}, \qquad (1.21)$$

$$E_x = T\frac{d}{dT}\left[\frac{\langle\tau\varepsilon\rangle}{eT\langle\tau\rangle} - \left(\frac{\varepsilon_F}{eT}\right)\right]\frac{dT}{dx} = -S\frac{dT}{dx}, \qquad (1.22)$$

where Eq. (1.22) applies because the Fermi level is independent of position in the absence of current. Further, the Seebeck coefficient, S, is defined in Eq. (1.22) as

$$S = -\frac{dV_x}{dT}, \qquad (1.23)$$

where V_x is the potential developed by the electric field E_x along the sample distance dx as a result of the thermal gradient dT.

Formally, the Seebeck coefficient may be expressed as

$$S = -\frac{T}{e}\frac{d}{dT}\left[\frac{\langle\tau\varepsilon\rangle}{T\langle\tau\rangle} - \frac{\varepsilon_F}{T}\right]. \qquad (1.24)$$

Generally for doped semiconductors in moderate electric fields, the Fermi energy is much greater than the kinetic energy of the electrons, so the second term of Eq. (1.24) will dominate the Seebeck coefficient. This coefficient is related to a quantity often described as the absolute thermoelectric power P where the Seebeck coefficient is related to P by the equation

$$S = T\frac{dP}{dT}. \qquad (1.25)$$

Thus the absolute thermopower is

$$P = \frac{\varepsilon_F \langle \tau \rangle - \langle \tau \varepsilon \rangle}{eT \langle \tau \rangle} = \frac{1}{eT} \left[\varepsilon_F - \frac{\langle \tau \varepsilon \rangle}{\langle \tau \rangle} \right]. \tag{1.26}$$

Thus, in a nondegeneraten-type semiconductor, the absolute thermopower, P, is given by Eq. (1.26) where the momentum relaxation time τ_e is frequently of the form $\tau_e = A\varepsilon^{-s}$ so that

$$\varepsilon_F \langle \tau_e \rangle / \langle \tau \rangle = (5/2 - s)kT \tag{1.27}$$

yielding

$$P = -\frac{k}{e} \left[(5/2 - s) - \frac{\varepsilon_F}{kT} \right]. \tag{1.28}$$

The sign of the Seebeck coefficient and the thermopower depends on the sign of the electronic carriers and will be negative for electrons and positive for holes. In semiconductors, there is the possibility of mixed conduction with both carriers present so the contribution to the thermopower is combined.

The thermopower of the nondegenerate semiconductor is significantly higher than the value characteristic of a classical electron gas where each particle has an energy of $3/2\, kT$ and so the thermoelectric power is approximately k/e equal to about $0.1\,\mathrm{mVK^{-1}}$.

In the case of a degenerate semiconductor or metal with $\tau_e = A\varepsilon^{-s}$, the absolute thermopower is given by

$$P = -\frac{k}{e} \left[\frac{\pi^2}{3}(3/2 - s) + \frac{kT}{\varepsilon_F} \right]. \tag{1.29}$$

The thermoelectric power for metals or degenerate semiconductors is smaller than that for conventional semiconductors by a factor of approximately kT/ε_F, or about 5×10^{-3}. Thus it is typical of most metals for the absolute thermopower to be $P \sim -0.03\, k/e \sim -3\,\mu\mathrm{VK^{-1}}$.

1.1.5 Electronic model

As a result of the form of the relationship for the absolute thermoelectric power of a degenerate electron gas that includes the ratio of kT/ε_F, the thermoelectric power of a metal or degenerate semi-conductor is orders of magnitude less than that of a nondegener-ate semiconductor. This is because in a material where the Fermi distribution applies, only a small fraction of the electrons, kT/ε_F, will actually contribute to the carrier transport, scattering processes or thermopower. The so-called Mott equation is frequently used to express the temperature dependence of the Seebeck coefficient of metals or degenerate semiconductors which decreases with decreasing temperature.[6]

That is,

$$S = \frac{\pi^2 k^2 T}{3e} \left[\frac{1}{\sigma(\varepsilon)} \frac{d\sigma}{d\varepsilon} \right]_{\varepsilon=\varepsilon_F}, \qquad (1.30)$$

where the conductivity $\sigma = n(\varepsilon)e\mu(\varepsilon)$ and the electron concentration is $n(\varepsilon) = D(\varepsilon)f(\varepsilon)$. Here $D(\varepsilon)$ is the density of states function and $f(\varepsilon)$ is the Fermi–Dirac distribution function. Using these yields

$$S = \frac{\pi^2 k^2 T}{3e} \left[\frac{1}{n} \frac{dn(\varepsilon)}{d\varepsilon} + \frac{1}{\mu} \frac{d\mu(\varepsilon)}{d\varepsilon} \right]_{\varepsilon=\varepsilon_F}. \qquad (1.31)$$

The magnitude of the Seebeck coefficient clearly depends on the type of thermoelectric material within the broad spectrum of thermoelec-tric materials that may include metals, nondegenerate, and degen-erate semiconductors. As noted earlier, the Seebeck coefficient is on the order of a few microvolts per degree kelvin for a typical metal as opposed to being on the order of a fraction of a millivolt per degree kelvin for a nondegenerate semiconductor. For semiconduc-tors, the doping of the semiconductor will determine the position of the Fermi level, and hence the magnitude of the thermoelectric power. If increased doping of an n-type semiconductor moves the Fermi level into the conduction band, it will cause the density of occupied electron states above and below the Fermi level to be more nearly equivalent thus reducing the thermoelectric power. Under these conditions, Eq. (1.31) will describe the Seebeck coefficient. The magnitude of the effect is dependent on form of the density of states which in a typical three-dimensional single-band crystal has an $\varepsilon^{1/2}$ energy dependence. Thus, an obvious method of improving the thermoelectric power of a material is to use a lower dimensional semiconductor whose electronic density will have a more nonunifom profile in energy space. Alternatively, the energy dependence of $\mu(\varepsilon)$ from energy-dependent carrier scattering such as that that may occur at a potential barrier (or hot-electron filter) will preferentially scatter low-energy electrons and cause an improved power factor in the material.

1.1.6 Figure of merit for thermoelectric materials

For nondegenerate extrinsic semiconductors with a single type of impurity, the Boltzmann equation can be used to evaluate the Seebeck coefficient, and the value of S is dependent on the position of $\varepsilon_{\mathbf{F}}$. If the semiconductor is intrinsic or compensated then both electrons and holes must used in determining the magnitude and sign of S. Therefore an intrinsic semiconductor with a significant quantity of both carriers will exhibit a lower value of S than its extrinsic counterpart. The value of S can vary significantly in an extrinsic semiconductor depending of the location of the Fermi level relative to the energy of the density of states function located within a band containing the majority carriers. Large variations in the density of states near the Fermi level may be used as a means to enhance the thermopower as reported by Heremans *et al.*[7]

Ultimately the figure of merit of a thermoelectric material is defined by the ability of the material to sustain a significant electric field between the hot and cold ends of the device while at the same time not experiencing a large decline of the thermal gradient between the two ends due to potentially large amounts of heat flow that would tend to reduce the gradient. Thus, the basic figure of merit of the thermoelectric system is defined as the dimensionless quantity ZT where

$$\mathrm{ZT} = \frac{\mathrm{S}^2\sigma T}{\kappa}, \tag{1.32}$$

where σ is the electrical conductivity of the thermoelectric material and κ is the thermal conductivity.[2] The thermal conductivity, κ, includes both the contributions from the electrons κ_e and the component arising from the phonons, κ_l, i.e., $\kappa = \kappa_e + \kappa_l$. For most semiconductors the larger component of the thermal conductivity is the phonon scattering except at very high temperatures or if the material is degenerate. Values of ZT on the order of ZT $\geq$ 1 are usually thought to constitute the threshold for a useful thermoelectric material. Values in this range are typically achieved by having a material that exhibits a large Seebeck coefficient, a high electrical conductivity and a low thermal conductivity. The ZT product may be improved by both increasing the Seebeck coefficient and increasing the electrical conductivity, and simultaneously reducing the thermal conductivity.

However, the thermal conductivity and the electrical conductivity are related through the Wiedemann–Franz ratio, where simple metals exhibit a ratio of $\kappa/\sigma = LT$. Here L is the Lorenz number $(\pi^2/3)(k/e)$, which is typically 2.45×10^{-8} V^{2}K^{-2}, and is a result of the fact that both the thermal and electrical currents are the result of the transport of the same fermions. This number can vary somewhat for some materials, and has been noted to be highly modified in one-dimensional systems.[8] It is often feasible to reduce the thermal conductivity in semiconductors associated with phonons by modifying the phonon scattering. A widely quoted axiom is that a good thermoelectric material is one that has the electronic properties of a high-quality crystal and the thermal properties of a glass.

In addition to ZT another related figure of merit is what is termed the power factor expressed by the product $S^2\sigma$. This quantity is solely dependent on the electronic properties of the material, but excludes the thermal conductivity that is determined by both electron and lattice scattering. Electron scattering is largely dominant in metals; however, the phonon scattering is the predominant form of scattering in semiconductors over most temperature ranges.[5] Temperature is also an influential factor in determining the relative importance of the two scattering mechanisms. The general approach in design of thermoelectric materials is to maximize the power factor by using a highly conducting semiconductor with a relatively narrow band gap and high carrier mobility to generate a high power factor or high ZT. The optimization of the figure of merit ZT, which depends

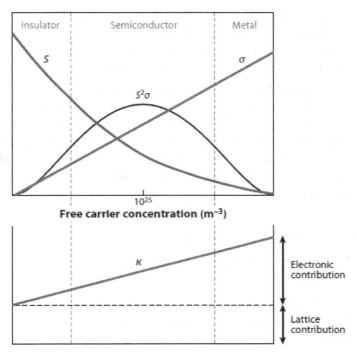

Figure 1.1. Dependence of the Seebeck coefficient (α), electrical conductivity (σ), thermal conductivity (κ), power factor ($\alpha^2\sigma$), and thermoelectric power (ZT) on carrier concentration. Used by permission from A. Shakouri, *ACS Ann. Rev. Mater. Res.*, **41**, 399–431 (2011).

on the product of the Seebeck coefficient and the electrical conductivity divided by the thermal conductivity, is illustrated in Fig. 1.1. This graphically demonstrates the complex relationship that must be achieved by a delicate balance between these interdependent quantities all of which are functions of the electron concentration. This ability to maximize ZT and its components is often described as carrier concentration tuning.

For homogeneous direct band gap materials there has been an analysis by Sofo and Mahan who suggest that there is an optimum energy band gap for semiconductors where phonon scattering is the principal mechanism limiting the thermal conductivity.[9] It is asserted that the optimum band gap is between $6kT$ and $10kT$, and appears to be consistent with the ZT of common thermoelectric materials. The analysis is based on the dependence of the transport parameters on the energy gap, and ultimately for direct gap materials is a function of the electron effective mass. In direct gap materials, this relation does not apply when the limiting scattering is due to ionized impurities, nor in indirect gap materials where the maximum ZT has no optimum value of band gap or effective mass. The optimization of band gap does not apply in inhomogeneous semiconductors, so it is feasible to avoid the limitations on scattering by introducing inhomogeneities that may include nanoscale particles or other inhomogeneous structures.

1.2 Thermoelectric Materials

1.2.1 Bulk thermoelectric materials

Throughout several decades beginning in the 1950s, thermoelectric effects were largely studied in bulk solids that were the class of materials exhibiting the best performance as thermoelectric devices. As noted above, the clear choice for thermoelectric power generation was a highly conducting semiconductor with a relatively narrow band gap of the order of 5–10 kT and high carrier mobility that is capable of producing a ZT product approaching one. A particular choice among various bulk binary semiconductors has been the compound Bi_2Te_3 that can be crystallized in a highly anisotropic layered structure with covalent bonds between the in-plane Bi and Te, and weak van der Waals bonding between adjacent Te layers. The compound $Bi_2\ Te_3$ has a narrow electronic energy band gap equal to approximately 0.16 eV with a hexagonal unit cell. Inasmuch as the material becomes compensated in the vicinity of normal ambient temperature, it is generally necessary to use alloys of Bi_2Te_3 with substituents such as Se, Te, or Sb. The addition of these constituents is used to fabricate thermoelectric power systems such as $Bi_2Te_{2.7}Se_{0.3}$ and $Bi_{0.5}Sb_{1.5}Te_3$ alloys with improved ZT values of approximately 1 as a result of the diminished thermal conductivity arising from the scattering of acoustic phonons in the alloy structures. For example, these materials have been shown to exhibit high values of electrical conductivity and a thermal conductivity of about $1.5\,\mathrm{Wm^{-1}K^{-1}}$ yielding a $ZT \sim 0.6$ at $300\,\mathrm{K}$.[10]

Another bulk system that has also been of interest as a thermoelectric material is the combination of Si and Ge in an alloy involving the two elements. Individually these two common semiconductors are not desirable candidates in a thermoelectric system because of their high lattice thermal conductivity on the order of $100\,\mathrm{Wm^{-1}K^{-1}}$; however when they are combined in a Si_xGe_{1-x} alloy the thermal conductivity is reduced to about $10\,\mathrm{Wm^{-1}K^{-1}}$ by phonon scattering with phonons and electrons.[11] The alloy system retains a high electronic carrier mobility so that the alloy displays a ZT product of approximately $ZT \sim 0.6$ in high-temperature ranges of technological interest for specialized applications.

1.2.2 Bulk approaches to high ZT thermoelectrics

1.2.2.1 *Solid-state chemistry model for high ZT*

The Bi_2Te_3 and Si_xTe_{1-x} alloy systems have been utilized widely in various thermoelectric systems for both power generation and solid-state refrigeration. There have been widespread efforts to enhance the performance of bulk systems beyond these basic systems. These solid-state chemistry methods to enhance the thermoelectric properties of the materials emphasize synthesis that attempts to lower the thermal conductivity by chemical or structural alterations of the basic thermoelectric systems.[12] There have been a number of efforts to extend work beyond the prototypical binary semimetallic Bi_2Te_3 system to ternary and quaternary systems with bulky atomic species exhibiting complicated structures with features such as carriers with high effective masses and lattices with poor thermal conductivity. Using this solid-state chemistry approach to materials design, layered structures such as $CsBi_4Te_6$ have been shown to produce ZT products of approximately 0.8 at $225\,\mathrm{K}$[13]; however, further improvements in these materials are necessary in order to reach usable thermoelectric parameters.

An additional compound of recent interest is a quaternary telluride that crystallizes with a PbTe structure having the relatively complex composition $AgPb_{10}SbTe_{12}$ where Bi may also be substituted for Sb. These compounds have demonstrated large values of $ZT \sim 1.7$ at temperatures near $700\,K$ as a consequence of their extremely low thermal conductivity. The low thermal conductivity is suggested to be the result of compositional modulation by nanocrystals at the 1–10 nm scale.[14]

Another attractive PbTe-based system has more recently been reported by Wu *et al.* who have synthesized a $PbTe_{0.7}S_{0.3}$ system doped by K that has achieved ZT values greater than 2.0 with an average ZT at least equal 2.0 over a temperature range from 673 to 923 K.[15] As reported, the use of K as a dopant of $PbTe_{0.7}S_{0.3}$ modifies the electrical conductivity by adjusting the carrier concentration, modifying the band structure, and yielding a grain structure including PbTe and PbS grains as well as other layers and precipitates. This effective combination of carrier concentration tuning, band structure modification, and reduced thermal conductivity due to increased nanoscale phases and defects results in an observed value of ZT equal to 2.2 at 923 K.

1.2.2.2 *"Half-Heusler" alloys*

Another materials system of interest are the "half-Heusler" alloys that are constructed from a MgAgAs structure having three interpenetrating fcc lattices somewhat reminiscent of the zinc blende GaAs structure that has two interpenetrating fcc lattices. The half-Heusler phase is a well-established intermetallic compound with the general formula XYZ, where X and Y are transition metals and Z is a metal or metalloid. In the half-Heusler material, the atoms X and Z populate a NaCl structure and Y is in one of two body diagonal positions while the other is not occupied. Thermal conductivity in these materials has been reduced through creating nanoscale phases in the host material to form nanocomposites, while in some recent reports this also appears to have increased the power factor, $S^2\sigma$. Half-Heusler alloys are doped by introducing atoms into the three occupied fcc lattices independently in order to modify the carrier concentration and produce defects that will reduce the thermal conductivity through phonon scattering. For example, in the TiNiSn system a ZT value ~ 1 is achieved after doping with high thermal stability up to a temperature of 900 K.[16]

1.2.2.3 *Skutterudites — rattlers*

A different approach to producing materials with low thermal conductivity is to synthesize crystal structures with large empty spaces that can be fully or partially filled with atoms that will perform as phonon scattering centers. The materials that are used in the void filling approach are often the so-called skutterudites that are a cubic structure consisting of eight corner octahedra (TX_6) where $CoSb_3$ is commonly used. Alternatives to Co are Ir or Rh, and for Sb are As or P. The octehedra create a large vacant region which is at the body centered position in the cubic lattice. This sizeable cavity has the capacity to hold substantial metal atoms. A wide variety of elements have been positioned in the holes including alkaline earth metals, lanthanides, and numerous Group IV elements. If the atoms in the empty spaces are heavy and relatively small, then the resultant disorder is greater, thus lessening the thermal conductivity associated with the lattice. A few doped skutterudites such as $In_x Co_4 Sb_{12}$ have demonstrated high ZT values.[17] Partially filled skutterudites whose voids are filled with electron donors exhibit low thermal conductivity and a high power factor relative to the mostly filled skutterudites. The rattling atom methodology as described in semiconducting $IrSb_3$ by Slack[18] has generated bulk thermoelectric materials with $ZT > 1$. The success of this method has encouraged widespread exploration of similar materials to achieve low thermal conductivity and high power factor.

1.2.2.4 *Chalcogenide compounds*

There are a number of chalcogenide compounds that are generally semiconductors, and have physical properties that make them promising as candidates for thermoelectric applications including energy gaps that are suitable as thermoelectric materials over a significant temperature range. Compounds such as $BaBiTe_3$ and

K_2Bi_8Se have weakly attached cations in channels exhibiting low lattice thermal conductivity. Some chalcogenides have been reported to have ZT values of 1–2.5. The solid-state synthesis of these innovative structures with greater anisotropy, low symmetry, and substantial unit cells has yielded low thermal conductivity helping to produce the high ZT values. Thallium chalcogenides including Tl_9BiTe_6 and the family of lead-based nanocomposite chalcogenides including materials such as $PbTe-AgSbTe_2$ are also reported to have ZT values of 2.1 at temperature near 800 K.[14] The latter materials appear to have compositional modulation at the nanometer scale that may influence both the carrier transport, as well as, the phonon transport being inhibited by nanocrystalline interfaces.

1.2.3 New materials — nanostructures, composites, and polymers

A recurring theme in many of the thermoelectric materials that show improved properties such as higher ZT product is the introduction of features that reduce thermal conductivity while maintaining good electrical conductivity. The phonon glass–electron crystal approach remains one of the most productive directions for exploring high-performance thermoelectric materials. Notwithstanding that crystalline semiconductors are the closest approximation to providing the compromise between the electronic aspects of the material, i.e., the electrical conductivity and the Seebeck coefficient, it is also clear that a low thermal conductivity as is present in an amorphous glassy material is the other requisite for an efficient thermoelectric material. As the majority of the contribution to the thermal conductivity in semiconductors often comes from the phonons, the phonon thermal conductivity can be diminished without a substantial change in the electrical conductivity by alloying as described earlier in the case of Bi_2Te_3 with Sb_2Te_3. Numerous complex crystal systems have been synthesized and investigated with the goal of attaining phonon glass/electron crystal materials in the search for high values of ZT. The new class of advanced bulk materials has been productive in generating materials of higher ZT and wider temperature ranges of application.[19]

1.2.3.1 *Nanostructures/composites*

Beyond these measures that are used to maximize the thermopower of bulk thermoelectric materials, another tactic being utilized is to use artificial structures that either modify the crystal periodicity or create a system whose dimensionality takes advantage of sharp features in the electron density of states. Confinement of the electronic carriers in a two-dimensional plane, a one-dimensional line or a zero-dimensional dot seeks to benefit from improved electronic properties of the candidate materials.[20] Moreover, these nanostructured materials potentially exhibit increased phonon scattering with a resultant lower lattice thermal conductivity. In addition to the benefits of nanostructures in thermoelectric materials with superlattices and quantum well materials, there is potentially an even greater improvement in the thermoelectric properties of nanocomposite materials. These materials entail the inclusion of nanostructured constituents[21] in a bulk material that may be either the same composition as the host bulk thermoelectric material, or may be an entirely different species. In either case, these modifications result in improved thermoelectric properties since the nanostructures and interfaces cause increased phonon scattering without having an equivalent degrading effect on the electronic properties.

As noted earlier, the maximum efficiency of thermoelectric energy conversion is expressed in terms of the figure of merit, ZT, that was given as

$$ZT = \frac{S^2 \sigma T}{\kappa},$$

which depends on the Seebeck coefficient, S, the high electrical conductivity, σ, and, κ, the thermal conductivity of the thermoelectric material. In order for the prospective thermoelectric material to possess a large figure of merit, it must have a large Seebeck coefficient, a high conductivity, and low thermal conductivity. However, the three characteristics are interrelated so that if we attempt to maximize S and minimize κ_e by using a low impurity concentration in a semiconductor it yields a material with low electrical conductivity. On the other hand, creating a high impurity concentration in the

semiconductor will generate a high electrical conductivity yet it has
the effect of decreasing the Seebeck coefficient, S, and increasing κ_e.

This conflicting dependence on various transport characteristics
has been a historic barrier to achieving improved thermoelectric
materials. Two methods used in bulk materials to reach ZT values approximately equal to 1 have depended on utilization of alloy
systems to increase phonon scattering and strategic optimization
of doping concentrations. However, a proposal to use materials of
reduced dimensionality by Hicks and Dresselhaus[20,21] provided a
seminal transformation to the approach to thermoelectric materials
design and synthesis. The predictions of significant improvements to
the figure of merit, ZT, in materials of lower dimensionality added a
new element to materials development that provided the prospect of
independently modifying electron and phonon transport much more
effectively than the prior tactics constrained to seeking the optimum
carrier concentration and a search for new thermoelectric materials.
As a result of successful strategies based on creating nanostructures
to reduce thermal conductivity, nanostructured and nanocomposite
materials have been synthesized whose figures of merit have reached
and exceeded values of ZT = 2. In particular, these nanostructured
materials provide the flexibility to separately adjust the transport
properties such as the electrical conductivity and the Seebeck coefficient. For example, note that the electrical conductivity, σ, is
given by

$$\sigma = -\frac{e^2}{3} \int v^2 \tau \frac{\partial f_0}{\partial \varepsilon} D(\varepsilon) d\varepsilon, \qquad (1.33)$$

while the Seebeck coefficient is

$$S = \frac{1}{e} \int \left(\frac{\varepsilon - \varepsilon_F}{T} \right) v^2 \tau \frac{\partial f_0}{\partial \varepsilon} D(\varepsilon) d\varepsilon \Big/ \int v^2 \tau \frac{\partial f_0}{\partial \varepsilon} D(\varepsilon) d\varepsilon. \qquad (1.34)$$

The transport properties are dependent on the carrier concentration,
the electrochemical potential that governs the window for the Fermi
function $-\partial f_0 / \partial \varepsilon$, the product of the group velocity, v, the density
of states $D(\varepsilon)$, and the relaxation time, τ. As the carrier concentration increases, both the electrical conductivity and the electronic

component of the thermal conductivity increase inasmuch as both parameters are directly dependent on n. Conversely, the Seebeck coefficient is reduced when electron concentration rises; hence with the two divergent dependences on electron concentration there must be an ideal magnitude of electron concentration to achieve the largest power factor and figure of merit, ZT. The underlying origin of this dependence is the position of the Fermi level and shape of the density of states that are, in turn, determined by whether the material is a lightly doped or heavily doped semiconductor or a metal. Figure 1.2 illustrates the energy-dependent mobility, $\mu(\varepsilon)$, the density of states, $D(\varepsilon)$, and the Fermi–Dirac distribution function, $f(\varepsilon)$, yielding an energy-dependent conductivity in a p-type semiconductor.

1.2.3.2 *Band dependence of thermoelectric power*

As noted in the previous discussion, the thermoelectric performance of a material will depend on the electronic dimensionality and band structure. In particular, Hicks and Dresselhaus[22,23] proposed thermoelectric materials that included either two-dimensional or one-dimensional structures would exhibit enhanced figure of merit and power factor. This improved performance was expected to occur because the density of states functions in these low dimensional

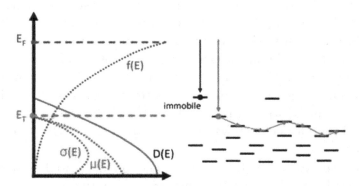

Figure 1.2. Energy-dependent mobility, $\mu(\varepsilon)$, density of states $(D(\varepsilon))$, Fermi–Dirac distribution function, $f(\varepsilon)$ showing energy-dependent conductivity, $\sigma(\varepsilon)$, μ in a p-type semiconductor. Used with permission from D.K Ko and C.B Murray, *ACS Nano*, **5**, 4810–4817 (2011).

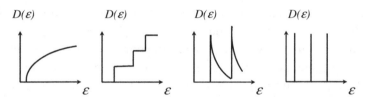

Figure 1.3. Density of states of electrons, $D(\varepsilon)$, for (a) 3D bulk crystal; (b) 2D quantum well; (c) 1D quantum wire; (d) 0D quantum dot.

structures have very sharply peaked forms in comparison to those exhibited by the more conventional three dimensional bulk materials. Figure 1.3 illustrates the densities of states of electrons, $D(\varepsilon)$, for 3D bulk crystals, 2D quantum wells, 1D quantum wires, and 0D quantum dots. Mahan and Sofo have projected that the optimum thermoelectric materials will have a density of states that is a delta function.[24] The narrow density of states in combination with a favorable positioning of the Fermi level provides a high power factor, and a significantly reduced electronic thermal conductivity.

The predictions by Hicks and Dresselhaus have been validated in many experiments that have followed this work. Specifically, two-dimensional quantum well structures in materials such as PbTe/PbS and Bi_2Te_3 superlattices have demonstrated large ZT enhancements beyond the limit of $ZT \sim 1$ that had persisted for more than a decade.[25,26] In three-dimensional materials, comparable distinct peaks in the density of states can be produced within an otherwise conventional band by introducing impurity states that are extended rather than being localized as is usually the case, and hence permit the electrons in the impurity states to contribute to charge transport. Impurities of this sort have been effective in generating improved thermoelectric power as demonstrated in PbTe doped with thallium.[27]

A logical expansion of the utilization of quantum wells and superlattices in thermoelectrics is to consider the quantum wires, quantum dots, and other forms of nanostructures as part of a composite thermoelectric material structure. Nanocomposites entail the inclusion of one or more of the types of nanostructures in a bulk material that may or may not have the same composition as the nanoinclusions. One of the most frequently used nanostructured components has been carbon nanotubes which are essentially a cylinder whose length is several orders of magnitude greater than its diameter, and is composed of a graphene sheet with a single layer of carbon atoms with a hexagonal crystal structure. This structure is called a single-walled nanotube (SWNT) in contrast to the so-called multiple walled nanotube (MWNT) that is composed of two or more concentric graphene layers. Carbon nanotubes alone exhibit a significant thermoelectric effect with observed values of the Seebeck coefficient S of SWNTs on the order of $45\,\mu VK^{-1}$ and MWNT of about $15\,\mu VK^{-1}$ in as-prepared films exposed to oxygen.[28,29] Exposure to oxygen has a pronounced effect on the Seebeck coefficient of nanotubes. The sign of S is positive for both of these types of CNTs indicative of the carriers being p-type.

Somewhat analogous to the graphene layers that are formed into carbon nanotubes, some other two-dimensional systems that have been fabricated as one-dimensional nanostructures such as the dichalogenides WS_2 and MoS_2, as well as layered structures like Bi_2Te_3. In the case of Bi_2Te_3, nanostructures have been formed by both electrode position and hydrothermal techniques. Zhao *et al.* have formed nanostructures Bi_2Te_3 using hydrothermal methods, and a variety of nanostructures of Bi_2Te_3 such as nanotubes and nanocapsules have been achieved, as well as alloys of Bi_2Te_3 with rare earth elements. Another interesting result is the formation of Bi_2Te_3 matrices with Bi_2Te_3 nanotubes as part of the composite structure.[30] Because of a low phonon contribution of $0.3\,Wm^{-1}\,K^{-1}$ to the thermal conductivity, the overall thermal conductivity is reduced to a value of $0.8\,Wm^{-1}K^{-1}$ in these nanocomposites, and a figure of merit greater than unity is achieved at about $T = 400\,K$.

Another approach to modifying the thermoelectric figure of merit has been demonstrated by Liu *et al.*[31] where the thermal conductivity of crystalline semiconductors has been modified in copper chalcogenides $Cu_{2-x}X$, where $X = S$, Se, or Te. These chalcogenides have highly intricate atomic configurations, and in the high temperature phase of this material the thermal conductivity associated with phonons is extremely low as a result of the arrangement of the copper atoms. In this regime, $Cu_{2-x}Se$ is a superionic conductor[32] where the Cu ions have high mobility in a system that has been described as a crystalline lattice of Se atoms surrounded by a "liquid" charged fluid of Cu ions. The fundamentally low lattice conductivity of this $Cu_{2-x}Se$ ultimately produces a high thermoelectric figure of merit ZT of approximately 1.5 at 1,000 K that ranks among the higher values observed for bulk materials. More importantly, the concept of using a material that contains a crystalline sublattice for electronic conduction that is surrounded by liquid-like ions provides an interesting direction for further investigation.

1.2.3.3 *Polymer/organic thermoelectrics*

Polymers are potentially intriguing materials for thermoelectric applications inasmuch as they have some characteristics that are not frequently associated with inorganic semiconductors including inexpensive processing at low temperature, mechanical flexibility, and adjustable electronic energy levels. Some solution processable organic precursors and quantum dots have been developed. Recent progress in synthesis of solution-processable polymer–inorganic hybrids with interesting thermoelectric properties is beginning to emerge in the literature. Among the most promising are polymer–nanocomposites where the second phase is either a metallic, inorganic, or carbon-based nanomaterial such as carbon nanotubes or graphene.

In pursuing polymer materials as an alternative to the complex inorganic compounds, alloys and composites it is important to recognize that these polymers have both decided advantages that include their low cost, less complex processing, and absence of scarce or harmful constituents. In comparison to inorganic thermoelectric materials, polymers typically have low thermal conductivity values that are generally of the order of $0.5\,Wm^{-1}K^{-1}$. Conversely, polymers typically have electron concentrations and electrical conductivities that are orders of magnitude lower than inorganic materials. Hence, to make polymer thermoelectric materials competitive with inorganics it is essential to modify the electron properties of the polymers to increase the electron concentration and conductivity sufficiently to bring the power factor of polymers to a competitive range.

1.2.3.3.1 *Electronic properties of polymers*

The electronic properties associated with organic semiconductors and polymers are determined by the structure of the monomers and how the monomer units are ordered in the solid-state polymer. A typical organic polymer is constructed of conjugated units with associated side chains that will determine the interactions in the units of the polymer. The customary energy band gap usually associated with an inorganic semiconductor compound can be described as the energy difference between the highest occupied molecular orbital and the lowest unoccupied molecular orbital states in the polymer. Thus the electronic energy levels of the polymer can be modified by changing the fundamental monomer unit or by adjusting the coupling between the monomer units. In simple polymers, the energy levels can be modified by doping while in more complex copolymers adding monomers with different energy levels results in the copolymer having an energy gap resulting from the energy level difference between the respective molecules.

Inasmuch as organic materials have bonding energies associated with weak van der Waals forces, the bonding is small as compared to typical inorganic semiconductors. Moreover, the electronic bands in the organic solids are relatively narrow compared to the band widths in the inorganic materials. Finally, synthesis of organic materials inherently yields materials with orders of magnitude higher impurity levels, as well as significantly higher defect density due to usually imperfect crystal structure. The fact that the energy level structure of the organic polymer will have a much higher defect and impurity density is responsible for broadening of impurity levels into impurity bands that overlap the edges of the energy bands thus creating band tails similar to those observed in disordered inorganic semiconductors. Ultimately, these bands are important in determining the transport characteristics of organic semiconductors by producing effective electron and hole mobilities that are frequently in the range

of 10–10^{-3} cm^2 V^{-1}s^{-1}. Transport in these models is often described
by either as hopping of carriers between electronic states, or a model
based on a mobility edge for carriers as has been used in disordered
inorganic semiconductors. Further, if the doping level of the polymer
needed to reach a high conductivity causes the position of the Fermi
level to move into the conduction band or the band tail, then the
number of states above and below the Fermi level will be nearly equal
thus creating a compensated material with a low Seebeck coefficient.
To overcome these difficulties in achieving the necessary electronic
properties, several approaches may be used including utilization of
structures with sharp peaks in the effective density of states such as
found in low-dimensional nanostructures, or by modifying the shape
of the density of states by creating a polymer blend of materials with
different carrier electronic energy levels.[33]

1.2.3.3.2 *Polymer thermoelectric materials*

An assortment of common conducting polymers have been
explored for their applicability as thermoelectric materials including polyanaline,[34] poly(thienothiophene),[35] poly(3,4-ethylenedioxy-
thiophene) (PEDOT),[36,37] ethylenedithiolate-metal coordination
polymers,[38] carbazole polymers,[39,40] poly(methoxyyphenyleneviny-
lenes),[41,42] poly(3-hexylthiophene),[43] polypypyrrole,[44,45] and others.
Figure 1.4 shows the structure of a number of the common molecules
that are the basis for thermoelectric polymers. An excellent review
article about polymer thermoelectric materials has summarized
many of their properties.[46] The majority of these polymers have
proven to have low Seebeck coefficients and low electron concentration with the exception of polyacetylene[47,48] that possesses high
electrical conductivity, but its properties are too unstable in air to be
utilized in thermoelectric applications. Some of the most successful
of recent efforts to improve polymer-based thermoelectric materials
has occurred in polymers where the carrier concentration has been
optimized, the morphology of the structure has been controlled, and
nanocomposites have been incorporated in the material.[49] Table 1.1
summarizes the thermoelectric properties of a number of polymer
and polymer-composite materials. A number of polymers have been

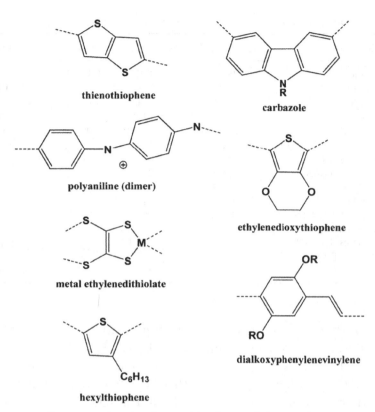

thienothiophene

carbazole

polyaniline (dimer)

ethylenedioxythiophene

metal ethylenedithiolate

dialkoxyphenylenevinylene

hexylthiophene

Figure 1.4. Structure of molecules used in thermoelectric polymers. From T.O. Poehler and H. E. Katz, *Energy. Environ. Sci.*, 5, 8110 (2012). Reproduced by permission of The Royal Society of Chemistry.

explored in depth for their thermoelectric properties, but poly(3,4-ethyylenedioxythiophene) complexed with polystyrene sulfonic acid (PEDOT-PSS) has been distinctive in displaying performance that approaches practical performance characteristics. However, a much wider range of materials have also shown promise that is based on polymer–composites particularly those involving the PEDOT-PSS family.

Bubnova *et al.*[37] demonstrated the importance of optimizing the carrier concentration in the conducting polymer, poly(3,4-ethyylenedioxythiophene) complexed with polystyrene sulfonic acid

Innovative Thermoelectric Materials

Table 1.1. Properties of polymer thermoelectric materials

Sample	$\sigma(\text{Scm}^{-1})$	$S(\mu VK^{-1})$	ZT	Ref.
Polyacetylene	28,500	1,077	1.1×10^{-2}	[47–49]
Polyanaline	320	225		[34, 46, 49]
Polypyrrole	340	7.14	3.2×10^{-4}	[44, 45, 49]
PEDOTnanowires	7–40	33–122		[55]
PEDOT:PSS/DMSO/EG	220–298	12.5–14.2	0.006–0.01	[36]
PEDOT:PSS/5%DMSO	298	12.65	0.01	[62]
PEDOT:PSS/EG	870	62	0.28	[51]
PEDOT:PSS/DMSO	960	73	0.4	[51]
PEDOT-Tos	~67	~220	0.25	[37]
PP-PEDOT	1,354	80–190	~1.02	[52]
Polymermatrix/PEDOT: PSS/SWCNT	1,050	25	0.02	[54]
PEDOT:PSS/Bi2Te3 (PH1000/FET)	50–250	50–150	0.08/	[57]
PEDOT:PSS/Te	19.3	163	0.1	[56]

(PEDOT-PSS). The observed thermoelectric figure of merit, ZT, of PEDOT-PSS was orders of magnitude lower than that of conventional Bi_2Te_3 alloys because an excess of nonionized dopant atoms diminish the conductivity of the PEDOT-PSS. When the PSS is replaced by tosylate (Tos), the resultant PEDOT-Tos may exhibit an electrical conductivity of more than $3 \times 10^2 \text{ Scm}^{-1}$ indicative of the fact that the nonionized dopant atom component is low. The high conductivity of PEDOT-Tos has been found to originate in crystalline nanodomains of PEDOT with Tos ions outside the PEDOT polymer chains. Increasing the carrier concentration in this instance has increased its power factor by more than an order of magnitude, and yielded a ZT = 0.25.

Bubnova *et al.*[50] have also established that polymers can be modified to have semimetallic properties as well as the semiconducting or insulating properties that we normally associate with polymeric materials. As is known, semimetals usually exhibit no direct bandgap and an extremely small electronic density of states at the Fermi level. Semimetal properties in polymers are obviously interesting inasmuch as inorganic thermoelectric materials of commercial interest generally are semimetals with higher Seebeck coefficients and electrical conductivity than metals.

The importance of minimizing the nonionized dopant concentration in polymer thermoelectric is a fundamental element in improving the transport properties of the material by increasing the carrier mobility and ultimately the power factor. By using PEDOT:PSS films doped with ethylene glycol(EG) or dimethylsulfoxide (DMSO) to improve conductivity, researchers have been able to deduce the optimum doping to maximize the thermoelectric power factor in PEDOT:PSS. They have achieved values of ZT derived from the power factor and thermal conductivity of $ZT = 0.42$ in PEDOT:PSS(DMSO) and $ZT = 0.25$ in PEDOT:PSS(EG).[51]

Highly conductive PEDOT films prepared by solution casting polymerization have demonstrated high power factors when the oxidation level is controlled electrochemically. In these materials, the electrical conductivity is decreased while the Seebeck coefficient is increased allowing the power factor to be maximized at a specific oxidation state. Use of electrochemical control of the oxidation state in a thin film thermoelectric material configuration has been demonstrated to generate improved thermoelectric performance in some cases compared to chemically controlled samples. PEDOT films synthesized with a combination of a pyridine base and a triblock copolymer PEPG[(poly(ethylene glycol)–poly(propylene glycol)–poly(ethylene glycol)] led to the preparation of a PEDOT film prepared using a mixture of pyridine and PEPG that is called PP-PEDOT. PP-PEDOT displays high conductivity ($1362 \, \mathrm{S \, cm^{-1}}$) and a maximum power factor of $1270 \, \mu\mathrm{Wm^{-1}K^{-2}}$. The ZT of PP-PEDOT has been estimated from the power factor and the thermal conductivity of PEDDOT to be approximately 1.02.[52]

1.2.3.4 *Polymer-composite thermoelectric materials*

An answer to the problem that has existed in polymer thermoelectric materials where there is a conflict between the requirement for high electrical conductivity, low thermal conductivity, and a high power factor is to use composite structures composed of a polymer and a nanostructured component in a thermoelectric material. This combination can combine the scattering properties of the nanomaterial to effectively scatter phonons associated with both the electronic carriers and the lattice yielding a low value of thermal conductivity. Conversely, the nanomaterial often contributes to a higher electrical conductivity in the composite than would be observed in the conjugated polymer. In addition to improving the conductivity of the polymer composite, the nanostructures may add a sharper contour to the density of states as well as generating additional selectivity to the energy transport. Considerable promise has been shown in using carbon nanotubes (CNTs) in polymer composites to enhance their thermoelectric properties. It has been demonstrated that in a composite composed of a polymer and CNTs that the CNTs provide a conducting path through the composite resulting in an electrical conductivity much higher than that of the underlying polymer although the CNTs only constitute a small percentage of the total volume. Further, it has been conjectured that the CNTs can induce the polymer chains to be more ordered as a result of π–π interactions between the polymer chains and the CNTs during polymerization that in turn increased the electron mobility.[53] However, the thermal conductivity is not markedly increased so the overall thermoelectric figure of merit is appreciably increased. A demonstration of this improvement was done by Kim *et al.*[54] where PEDOT-PSS was used in a nanocomposite morphology together with CNTs that yielded a large improvement in ZT over pure PEDOT-PSS. A polymer consisting of nanowires of the PEDOT itself has been reported to improve the thermoelectric properties above that of a PEDOT film for many of the same reasons as observed for a polymer–nanocomposite combination.[55]

An alternative to using carbon nanotubes in a polymer composite is introduction of other semiconductor nanostructures in a polymer blend to decrease the thermal conductivity and to enhance processing of the material.[28,29] See *et al.* introduced Te nanorods in PEDOT-PSS during synthesis from solution casting of a film with a network of nanodimensional organic/inorganic interfaces where the thermal conductivity was reduced, but the electrical conductivity improved.[56] These composites had thermoelectric properties with a Seebeck coefficient of $163 \, \mu VK^{-1}$ and had a power factor of $70 \, \mu WmK^{-2}$. In several instances various inorganic semiconductors have been incorporated in the composite to generate a large power factor. For example, Zhang *et al.*[57] included Bi_2Te_3 as a component in PEDOT-PSS where the holes were the majority electronic carrier, and the material exhibited a significant improvement in the observed power factor above that in a pure PEDOT polymer. The resulting value of Seebeck coefficient in an n-type composites was approximately $80 \, \mu VK^{-1}$ and had a power factor of $47 \, \mu WmK^{-2}$.

Likewise, Toshima *et al.* used polyaniline to produce a similar increased power factor in a hybrid of polyaniline and Bi_2Te_3.[58] When Bi_2Te_3 nanoparticles were added to a polyaniline solution and solution cast on a glass substrate, the resultant films showed better thermoelectric performance than the pure organic conducting polymer. A hybrid film prepared from a physical mixture of polyaniline and Bi_2Te_3 nanoparticles in a cast layer had a power factor of about 50–$80 \, \mu WmK^{-2}$ that compares favorably to that of a pure polyaniline film display a power factor of about $1 \, \mu WmK^{-2}$ at a temperature of $350 \, K$. Thus, the polymer–nanotube composite prepared by either physical or solution mixing exhibited improved Seebeck coefficients compared to the individual components. The hybrid films fabricated by physical mixing displayed values of Seebeck coefficients generally higher by one order of magnitude than those of pure polyaniline films.

In some cases, the composite material has performed better than either of the components of the composite which seemingly appears to be in conflict with Bergmann's hypothesis that states that the ZT of the composite cannot exceed the ZT of the individual constituents.[59] As a case in point is the report by See *et al.*[56] where using a solution processing method the authors synthesized a composite of PEDOT-PSS with Te nanowires and the composite had a ZT that exceeded that of either of the components. It has been postulated by See *et al.* that the conventional limits on ZT result from an effective medium averaging process in determining ZT that is no longer applicable when the length scale of the phenomena corresponding to the observed values is on the order of the length scale of the constituents of the composites.[56] The experiments also demonstrated that the transport in the nanoscale organic/inorganic structures could create combined thermal and electronic transport capable of significant values of ZT (>0.1). Subsequent work reveals that carrier transport in the complex organic/inorganic structures involves interfacial effects that are beneficial in optimizing ZT.[60] Recent measurements of the transport at high frequencies indicate that the complex composite materials have more de-localized electronic carriers that are critical in increasing the electrical conductivity, and hence improving the thermoelectric figure of merit.[61] The combined results of this work appear to provide a path toward enhanced transport and thermoelectric properties in this significant group of complex materials.

1.3 Summary

Organic semiconductors and composites are clearly demonstrating significant promise as future replacements for some inorganic thermoelectric materials predominantly for moderate temperature ranges. Specific polymer families such as those incorporating PEDOT have demonstrated notable power factors and figures of merit with ZT approaching 1.0. The application of nanocomposites to reduce thermal conductivity in polymer nanocomposite structures has been responsible for a notable improvement in power factor of these nanomaterials as thermoelectric elements. Several important strategies such as band structure modification in low-dimensional materials with selective densities of states, selectively increased carrier scattering, and control of dopant concentrations are proving to be effective in improving the thermoelectric properties of these materials. Exciting prospects for new semiconducting compounds and polymers with improved electronic properties provide expectations of further improvement in the thermoelectric figures of merit. The degrees of freedom and complexity of materials incorporating various combinations of polymers, reduced dimensionality, and composite structures provide immense promise of a continuation of improvement for this generation of new thermoelectric materials. In this volume, the contributors have provided a series of chapters that are intended to highlight many of the challenges, opportunities, and future prospects in this exciting field. There are eight chapters that basically encompass subjects that describe the directions where research is occurring that is intended to expand and improve the materials that will be obtainable for future applications in thermoelectric power generation.

References

1. A. F. Ioffe, *Semiconductor Thermoelements and Thermoelectric Cooling*, Infosearch: London, (1957).
2. D. M. Rowe, ed. *CRC Thermoelectrics Handbook*, CRC Press: Boca Raton, FL, (2006).
3. L. E. Bell, Cooling, heating, generating power and recovering waste heat with thermoelectric systems. *Science*, **321**(5895), 1457–1461 (2008).
4. D. K. Ferry, *Semiconductors*, Macmillan: New York (1961).
5. R. A. Smith, *Wave Mechanics of Crystalline Solids*, Chapman and Hall, and John Wiley and Sons: London and Hoboken (1961).
6. N. F. Mott and E. A. Davis, *Electronic Processes in Non-crystalline Materials*, Clarendon Press: Oxford (1971), p. 47.
7. J. P. Heremans, V. Jovivic, and E. S. Toberer, Enhancement of the efficiency in PbTe by distortion of the electronic density of states. *Science*, **321**, 554–557 (2008).
8. N. Wakeham, A. F. Bangura, X. F. Xu, J. F. Mercure, M. Greenblat, and N. E. Hussey, Gross violation of the Wiedemann–Franz law in a quasi-one-dimensional conductor. *Nat. Commun.*, **2**, 396 (2011).
9. J. O. Sofo and C. D. Mahan, Optimum band gap of a thermoelectric material. *Phys. Rev. B*, **49**, 4565–4570 (1994).
10. H. Scherrer and S. Scherrer, Thermoelectric Properties of Bismuth Antimony Telluride Solid Solutions D. M. Rowe, ed. *CRC Thermoelectrics Handbook*, 27-1–27-18. CRC Press: Boca Raton, FL (2006).
11. C. Wood, Materials for thermoelectric energy conversion. *Rep. Progr. Phys.*, **51**, 459–539 (1988).
12. M. G. Kanatzidis, Structural evolution and phase homologies for "design" and prediction of solid-state materials. *Acc. Chem. Res.*, **38**, 361–370 (2004).
13. D. Y. Chung, T. Hogan, P. Braziz, M. Rocci-Lane, C. R. Kannerwurf, M. Bastea, C. Uher, and M. G. Kanatzidis, $CsBi_4Te_6$: A high performance thermoelectric material for low-temperature applications. *J. Am. Chem. Soc.*, **287**, 1024 (2000).

14. K. F. Hsu, S. Loo, F. Guo, W. Chen, J. S. Dyck, U. Uher, T. Hogan, E. K. Polychroniadis, and M. G. Kanadzidis, Cubic $AgPb_mSbTe_{2+m}$ bulk thermoelectric materials with high figure of merit. *Science*, **303**, 818 (2004).
15. H. J. Wu, L. D. Zhao, F. S. Zheng, D. Wu, Y. L. Pei, X. Tong, M. G. Kanatzidis, and J. G. He. Broad temperature plateau for thermoelectric figure of merit $ZT > 2$ in phase separated $PbTe_{0.7}S_{0.3}$. *Nature Comm.*, **5**, 4515 (2014). doi:10.1038/ncomms 5515.
16. W. Xie, A. Weidenkaff, X. Tang, Q. Zhang, J. Poon, and T. M. Tritt, Recent advances in nanostructured thermoelectric half-Heusler alloys. *Nanomaterials*, **2**, 379–412 (2012).
17. T. He, J. Chen, H. D. Rosenfeld, and M. A. Subramanian, Thermoelectric properties of indium-filled skutterites. *Chem. Mater.*, **18**(3), 759–762 (2006).
18. G. A. Slack and V. G. Tsoukala, Some properties of semiconducting $IrSb_3$. *J. Appl. Phys.*, **76**, 1665 (1994).
19. G. S. Nolas, J. Poon, and M. Kanatzidis, Recent developments in bulk thermoelectric materials. *MRS Bulletin*, **31**, 198–205 (2006).
20. L. D. Hicks and M. S. Dresselhaus, Effect of quantum well structures on the thermoelectric figure of merit. *Phys. Rev. B*, **47**, 12727–12131 (1993).
21. M. S. Dresselhaus, G. Chen, M. Y. Tang, R. Yang, H. Lee, D. Wang, Z. Ren, J. Fleurial, and P. Gogna, New directionsfor low-dimensional thermoelectric materials. *Adv. Mater.*, **18**, 1043–1053 (2007).
22. L. D. Hicks and M. S. Dresselhaus, Thermoelectric figure of merit of a one-dimensional conductor. *Phys. Rev. B* **47**, 16631(1993).
23. L. D. Hicks, T. C. Harman, X. Sun, and M. S. Dresselhaus, Experimental study of the effect of quantum-well structures on the thermoelectric figure of merit. *Phys. Rev. B*, **53**, R10493 (1996).
24. G. Mahan and J. Sofo, The best thermoelectrics, *Proc. Natl Acad. Sci.*, **93**(15), 7436–7439 (1996).
25. T. C. Harman, P. Taylor, M. Walsh, and B. E. LaForge, Quantum dot superlattice thermoelectric materials and devices. *Science*, 297, 2229 (2002).
26. R. Venkatsubramanian, R. Sivola, T. Colpitts, B. and O'Quinn, Thin film thermoelectric devices with high room temperature figures of merit. *Nature*, **13**, 597–602 (2001).
27. J. Heremans, V. Jovovic, E. Toberer, A. Saramat, K. Kurosaki, A. Charoenphakdee, S. Yamanaka, and G. Snyder, Enhancement of thermoelectric efficiency in PbTe by distortion of the electronic density of states. *Science*, **321**, 554 (2008).

28. P. G. Collins, K. Bradley, N. Ishigami, and A. Zettl, Extreme oxygen sensitivity of electronic properties of carbon nanotubes. *Science*, **287**, 1801 (2000).
29. P. K. Bradley, S. H. Jhi, P. G. Collins, J. Hone, M. L. Cone, and A. Zettl, Is the intrinsic thermoelectric power of carbon nanotubes positive? *Phys. Rev. Lett.*, **85**, 4361 (2000).
30. X. B. Zhao, X. H. Ji, Y. H. Zhang, T. J. Zhu, J. P. Tu, and Z. B. Zhang, Bismuth telluridenanotubes and the effect on the thermoelectric properties of nanotube-containing nanocomposites. *Appl. Phys. Lett.*, **86**, 062111 (2005).
31. H. Liu, X. Ahi, F. Xu, L. Zhang, W. Zhang, L. Chen, L. Qiang, C. Uher, T. Day, and G. J. Snyder, Copper ion liquid-like thermoelectric, *Nature Mater.*, **11**, 422 0 425 (2012).
32. J. B. Boyce and B. A. Huberman, Superionic conductors: Transitions, structures, dynamics. *Phys. Rev.*, **51**, 189–265 (1979).
33. J. Sun, M.-L. Yeh, B. J. Jung, B. Zhang, J.Feser, A. Majundar, and H. E. Katz, Simultaneous increase in the Seebeck coefficient and conductivity in a doped poly(alkylthiophene) blend with defined density of states. *Macromolecules*, **43**, 2897–2903 (2010).
34. N. Mateeva, H. Niculescu, J. Schlenoff, and L. R. Testardi, Correlation of Seebeck coefficient andelectrical conductivity in polyanaline and polypyrrole. *J. Appl. Phys.*, **83**, 3111–3117 (1998).
35. R. R. Yue, S. A. Chen, B. Y. Lu, C. C. Liu, and J. K. Xu, Facile electrosynthesis and thermoelectric performance of free-standing polythieno[3,2-b] thiophene films. *J. Solid State Electrochem.*, **15**, 539–548 (2011).
36. C. C. Liu, B. Y. Lu, J. Yan, J. K. Xu, R. R. Yue, Z. J. Zhu, S. Y. Zhou, X. J. Hu, Z. Zhang, and P. Chen, Highly conducting free standing poly(3,4-ethylenedioxythiophene)/poly(stryrenesulfonate) films with improved thermoelectric performances. *Synth. Met.*, **160**, 2481–2485 (2010).
37. O. Bubnova, Z. U. Khan, A. Malti, S. Braun, M. Fahlman, M. Berggren, and X. Crispin, Optimization of the thermoelectric figure of merit in the conducting polymer poly(3,4 -ethylenedioxythiophene) *Nature Mater.*, **10**, 429–433 (2011).
38. Y. M. Sun, P. Sheng, C. A. Di, F. Jiao, W. Xu, D. Qiu, and D. B. Zhu, Organic thermoelectric materials and devices based on p and n type poly(metal 1,1,2,2-ethenetetrathiolate)s. *Adv. Mater.*, **24**, 932–937 (2012).
39. R. B. Aich, N. Blouin, A. Bouchard, and M. Leclerc, Electrical and thermoelectric properties of poly(2,7-carbazole) derivatives, *Chem. Mater.*, **21**, 751–757 (2009).

40. I. Levesque, P. O. Bertrand, N. Blouin, M. Leclerc, S. Zecchin, G. Zotti, C. I. Ratcliffe, D. D. Klug, X. Gao, F. M. Gao, and J. S. Tse, Synthesis and thermoelectric properties of polycarbazole polyindolocarbazole, and polyindolocarbazole derivatives. *Chem. Mater.*, **19**, 2128–2138 (2007).

41. Y. Hiroshige, M. Ookawa, and N. Toshima, High thermoelectric performance of poly(2,5-dimethoxyphenylenevinylene) and its derivatives. *Synth. Met.*, **156**, 1341–1347 (2006).

42. Y. Hiroshige, M. Ookawa, and N. Toshima, Thermoelectric figure-of-merit of iodine-doped copolymer phenylenevinylene with dialkoxyphenylenevinylen. *Synth. Met.*, **157**, 467–474 (2007).

43. Y. Xuan, X. Liu, S. Desbief, P. Leclerc, M. Fahlman, R. Lazzaroni, M. Berggren, J. Cornil, D. Emin, and X. Crispin, Thermoelectric propeties of conducting polymers: The case of poly(3-hexylthiophene). *Phys. Rev. B*, **82**, 115454 (2010).

44. K. Sato, M. Yamamura, T. Hagiwara, K. Murata, and M. Tolkumoto, Study on the electrical conduction mechanisms of polypyrrole films. *Synth. Met.*, **40**, 35–48 (1991).

45. W. Y. Lee, Y. W. Park, and Y. S Choi, Metallic electric transport of PF6-doped polypyrrole: DC conductivity and thermoelectric power. *Synth. Met.*, **84**, 341–342 (1997).

46. N. Dubey and M. Leclerc, *Conducting* polymers: Efficient thermoelectric materials. *J. Polym. Sci. Part B — Polym. Phys.*, **49**, 467–475 (2011).

47. Y. Nogami, H. Kaneko, and T. Ishiguro, On the metallic states in highly conducting iodine-doped polyacetylene. *Solid State Comun.*, **76**, 583–586 (1990).

48. O. Yoon, Y. W. Park, H. Shirakawa, and K. Akagi, Thermoelectric power and conductivity of the stretch oriented polyacetylene film doped with $MOCl_5$. *Synth. Met.*, **41**, 125–128 (1991).

49. Y. Du, S. Z. Shen, K. Cai, and P. S. Casey, Research progress on polymer-inorganic thermoelectric nano-composite materials. *Progress Poly. Sci.* **37**, 820–841 (2012).

50. O. Bubnova, Z. U. Khan, H. Wang *et al.*, Semi-metallic polymers. *Nature Mater.*, **13**, 190–194 (2014).

51. G.-H. Kim, L. Shao, K. Zhang, and K. P. Pipe, Engineered doping of organic semiconductors for enhanced thermoelectric efficiency. *Nature Mater.*, **12**, 719–723 (2013).

52. T. Park, C. Park, B. Kim, H. Shin, and E. Kim, Flexible PEDOT electrodes with large thermoelectric power factors to generate electricity by the touch of the fingertips, *Energy Environ. Sci.*, **6**, 788–792 (2013).

53. Q. Yao, L. Chen, W. Zhang, S. Liufu, and X. Chen, Enhanced thermoelectric performance of single–walled carbon nanotubes/polyanaline hybrid nanocomposites, *ACS Nano*, **4**, 2445–2451 (2010).

54. D. Kim, Y. Kim, K. Choi, J. C. Grunlan, and C. Yu, Improved thermoelectric behavior of nanotube-filled polymer composites with poly(3,4-ethylenedioxythiophene)poly(styrenesulfonate). *ACS Nano*, **1**, 513–523 (2010).

55. D. K. Taggart, Y. Yang, S. C. Kung, T. M. McIntire, and R. M. Penner, Enhanced thermoelectric metrics in ultra-long electrodeposited PEDOT nanowires. *Nano Lett.*, **11**, 125 (2011).

56. K. C. See, J. P. Feser, C. E. Chen, A. Majundar, J. J. Urban, and R. A. Segalman, Water-processable polymer–nanocrystal hybrids for thermoelectric, *Nanoletters*, **19**, 4664–4667 (2010).

57. B. Zhang, J. Sun, H. E. Katz, F. Fang, and R. L. Opila, Promising thermoelectric properties of commercial PEDOT-PSS materials and their Bi_2Te_3 powder composites. *Appl. Mater. Interfaces*, **2**, 3170–3178 (2010).

58. N. Toshima, M. Imai, and S. Ichikawa, Organic–inorganic nanohybrids as novel thermoelectric materials: Hybrids of polyanaline and bismuth(III) telluride nanoparticles, *J. Elec. Mat.*, **40**, 898–902 (2011).

59. D. J. Bergman, and O. Levy, Thermoelectric properties of a composite medium, *J. Appl. Phys.*, **70**(11), 6821–6833 (1991).

60. N. E. Coates, S. K. Yee, B. McCulloch, K. C. See, A. Majumdar, R. A. Segalman, and J. J. Urban, Effect of interfacial properties on polymer-nanocrystal thermoelectric transport. *Adv. Mater.*, **25**, 1629–1633 (1991).

61. J. N. Heyman, B. A. Alebachew, Z. S. Kaminski, M. D. Nguyen, N. E. Coates, and J. J. Urban, Terahertz and infrared transmission of an organic/inorganic hybrid thermoelectric material, *Appl. Phys. Lett.*, **104**, 141912 (2014).

62. K. Chang, M. Jeng, C. Yang, Y. Chou, S. Wu, M. Thomas, and Y. Peng, The thermoelectric performance of poly(3,4-ethylenedioxythiophene)/poly(stryrenesulfonate). *J. Elect. Mater.*, **38**(7), 1182–1584 (2009).

Chapter 2

Solution-Processable Molecular and Polymer Semiconductors for Thermoelectrics

Ruth Schlitz, Anne Glaudell and Michael Chabinyc

2.1 Introduction

Thermoelectric materials can convert a thermal gradient to an electrical potential or, by reciprocity, a potential difference driving an electrical current to a thermal gradient.[1] The basis of this ability is the Seebeck effect where a thermal gradient causes the diffusion of charge carriers leading to the generation of an electrical potential, $\alpha = -\Delta V / \Delta T$, where α is the thermopower also known as the Seebeck coefficient (μVK^{-1}). The thermopower of a material is a measure of the entropy per carrier due to its relationship with the electronic density of states (DOS). As the carrier concentration increases, the electrical conductivity σ increases, while the Seebeck coefficient decreases (Fig. 2.1).[2] In most materials, the thermal conductivity rises as a function of carrier concentration due to the electronic contribution surpassing the lattice conductivity. The dimensionless figure of merit for thermoelectric performance, $ZT = \alpha^2 \sigma T / \kappa$, relates to the ability of a material to be an efficient thermoelectric material, by comparing its ability to produce electrical power (thermopower $\alpha^2 \sigma$) to its ability to conduct heat via the thermal conductivity (κ). ZT maximizes at a particular carrier density

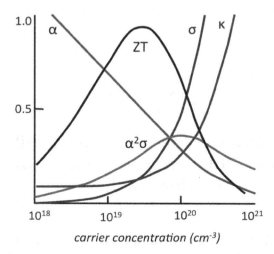

Figure 2.1. Schematic dependence of the thermopower (α), electrical conductivity (σ), and thermal conductivity (κ) as a function of carrier concentration expected for a single band semiconducting material. The left axis shows ZT which peaks for many inorganic semiconductors at carrier concentrations between 10^{19} and 10^{20} cm^{-3}. This figure is adapted from Ref. 2.

(and thereby a particular electrical conductivity) due to the interrelationship of the electrical and thermal properties of thermoelectric materials via carrier concentration.

Organic semiconductors have been studied extensively for applications in conductors and thin film electronics. High-performance organic photovoltaics (OPVs) with power conversion efficiencies near 10%,[3] all-printed circuits with both p- and n-type thin film transistors (TFTs),[4] and highly efficient organic light-emitting diodes (OLEDs)[5] have been demonstrated. These applications have driven the synthesis of a large range of materials with varying electronic structures[6–9] and an extensive study of the relationship of processing methods to their electrical performance.[10,11] The ability to process organic semiconductors using simple methods such as ink-jet printing make them attractive for the fabrication of thermoelectric devices.

Organic materials have not been widely explored as thermoelectric materials despite promising physical characteristics. Organic semiconducting materials have relatively low thermal conductivity

($\sim$0.3 $Wm^{-1}K^{-1}$), yet can have relatively high electrical conductivity (1–1,000 Scm^{-1}).[12] This combination of properties suggests promise as thermoelectrics near room temperature ($<$$\sim$200°C). A key challenge for the development of organic thermoelectrics is to develop the means to dope materials to sufficient charge-carrier densities to obtain high electrical conductivity and high thermopower. Additionally, traditional thermoelectric devices consist of p- and n-type bulk leg pairs, wired electrically in series and thermally in parallel, allowing large power outputs and temperature gradients (Fig. 2.1). Although one can make a single-carrier device, the most efficient devices of traditional geometry require both p- and n-type legs with similar thermoelectric properties.[2] While there are many known stable p-doped organic materials, there are very few n-type materials currently.

Both organic semiconducting molecules and polymers have been studied as thermoelectrics. Foundational work on single co-crystals of electron donor and acceptors revealed that organic semiconductors can have metallic behavior, high electrical conductivity ($\sim$1–10 Scm^{-1}) and signatures of 1D transport.[13–15] Such materials have exciting physical and electrical behavior, but they are challenging to process into devices due to their solubility and the difficulty of tuning their carrier concentration because of the need to maintain fixed stoichiometry between the donor and acceptor. Electrically doped semiconducting polymers provide a pathway toward greater processability because of the ability to tune the carrier concentration.[16] Iodine-doped polyacetylene has very high electrical conductivity ($>$1,000 Scm^{-1}) and thermoelectric performance rivals that of bismuth telluride at room temperature.[17] However, iodine-doped polyacetylene is known to be unstable.[18] Recent improvements in the synthetic chemistry of semiconducting polymers and stable doped polymers such as those based on 3,4-ethylenedioxythiophene (PEDOTs) suggest that these problems may be overcome.[19]

In this chapter, we focus on the thermoelectric behavior and transport models used to understand organic polymers and solution-processable organic semiconductors. The thermopower of vapor deposited materials has also been studied mainly in the context of injection layers for OLEDs and has been previously reviewed.[20] We present here an overview of recent progress in both p- and n-type materials and outline challenges for the future development of organic thermoelectrics.

2.2 Electronic Structure and Morphology of Organic Semiconductors

2.2.1 Electronic levels

The electronic properties of organic semiconductors are defined by both their molecular structure and organization in the solid state. The molecular structure of solution-processable organic semiconductors comprises conjugated units along with nonconjugated sidechains to improve solubility and to guide intermolecular interactions in the solid state (Table 2.1). Both molecular and polymeric semiconducting polymers can be processed from the solution, but the latter have been studied more extensively as thermoelectrics. The

Table 2.1. Structure of materials examined as p-type thermoelectrics and representative thermoelectric performance near 298 K.

Material	Dopant or counterion	σ (S/cm)	α (μV/K)	P.F. μWm^{-1}K^{-1}	Ref.
polyacetyelene (PA)	I_2	30,000	20	1,200	52
P3HT	NOPF$_6$	2	25	0.1	56
	Fe(triflimide)	180	31	30	57
PEDOT	complexed with PSS	880	73	470	63
	Tos	80	200	324	61

(*Continued*)

Table 2.1. (*Continued*)

Material	Dopant or counterion	σ (S/cm)	α (μV/K)	P.F. μWm^{-1}K^{-1}	Ref.
PEDOS-C6	electrochemical	335	31	354	65
PBTTT	Fe(triflimide)	220	13	3.7	59
	NOPF$_6$	54	13	0.9	60
PSBTBT	Fe(triflimide)	21	24	1.2	59
PCDTBT	FeCl$_3$	160	34	19	58
PCDTB	FeCl$_3$	87	40	14	58
	Fe(triflimide)	62	45	12	59

electronic gap between the valence (ionization energy (IE)) and conduction states (electron affinity (EA)) are frequently discussed by reference to molecular states, e.g., highest occupied molecular orbital (HOMO) and lowest unoccupied molecular orbital (LUMO), even in the solid state. Due to the difficulty of directly measuring the energetics of unoccupied electronic states, the conduction level (LUMO) is frequently estimated by the optical gap creating significant uncertainty due to the excitonic nature of the photoexcitations in organic materials.[21]

The electronic levels of semiconducting polymers can be tuned by the basic repeat unit structure and by interactions between repeat units (Table 2.1). Homopolymers such as poly(alkylthiophenes) or poly(acetylene) can be tuned by substituents, e.g., the addition of electronegative groups such as fluorine will lower the conduction levels. In co-polymers, the use of monomers with electronic offsets in the molecular states leads to the so-called donor–acceptor polymers where the electronic gap is defined by the interaction between repeat units. The donor–acceptor approach is widely used for low-gap polymers ($<\sim 1.7$ eV). Tuning of these levels is critical for the electrical doping and stability of the carriers to chemical reactions with water and oxygen. For example, a LUMO near -4.0 eV is found to lead to stability in the ambient for n-type materials and a HOMO below -5.0 eV lends stability for p-type materials.

Doping of organic semiconductors requires reagents that chemically oxidize or reduce the material to form charge carriers. For example, for p-type doping, electron acceptors with large EA, such as F_4TCNQ, can drive charge transfer with electron donors.[20] Alternatively, chemical oxidants such as iron salts or $NOPF_6$ are commonly used.[22] This process dramatically changes the solubility of organic semiconductors from that of the neutral form and in many cases dictates the level of doping that may be achieved by directly adding the dopant to the casting solution. Alternatively, an organic semiconducting film can be soaked in a nonsolvent containing the oxidant or reductant to form charge carriers, but may be problematic for doping thick films unless the diffusion of the reagent is rapid through the polymer. All of these processes cause significant changes to microstructure because of the incorporation of the dopant that perturbs the initial molecular packing of the host.

2.2.2 Morphology

Solution-processed organic semiconductors can form amorphous or polycrystalline films.[23,24] The electrical properties of organic thin films are found to be highly anisotropic due to their molecular ordering. The origin of the anisotropy is the observation that crystalline domains typically have a preferred texture (orientation) relative to the substrate.[24] Additionally, the size of ordered domains in thin films determines the density of boundaries that carriers must cross for transport. Due to a wide variety of molecular packing motifs in molecular materials, it is difficult to generalize about the role of grain boundaries on transport.[25] Domain boundaries do have a significant impact on charge transport, in general, and are likely to contribute to the thermopower in molecular materials.[26]

Semiconducting polymers may form semicrystalline films that are disordered due to the distribution of molecular weights and kinetic limitations during the deposition process. The molecular packing structure of most semiconducting polymers comprises extended polymer chains in parallel with a packing structure that is dependent on the structure of the sidechains. Polymers with linear sidechains tend to form lamellar structures, while those with branched sidechains may pack in a nearly hexagonal manner.[10] In both cases, conduction along the polymer chain is expected to be simpler due to the strong electronic coupling of the covalently bonded chain, smaller in the direction of chain-to-chain co-facial interactions and exceptionally poor in the direction of the sidechain packing. Polymer chains in films are anisotropic in both semicrystalline and more disordered semiconducting polymers. The molecular packing structure of most semiconducting polymers comprises extended polymer chains in parallel with a packing structure that is dependent on the structure of the sidechains (Fig. 2.2). Comparisons of molecular spectroscopies and both X-ray and electron scattering show that, in general, the chains are aligned along the substrate and the conjugated plane may be either edge-on or face-on to the substrate (Fig. 2.3).[10] In bulk samples formed by drop-casting methods, there may be anisotropies similar to thin films, but in these cases chain alignment is typically achieved by mechanical methods, e.g., drawing fibers or stretching films. Due to the possibilities of significant morphological anisotropies, it is important to examine both thermal and electrical transport properties in the same

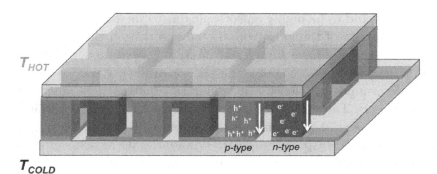

Figure 2.2. Schematic of a typical thermoelectric module with p-type and n-type legs connected electrically in series and thermally in parallel. In a thermal gradient, free carriers at higher temperature (T_{HOT}) will diffuse faster than those at lower temperature (T_{COLD}), resulting in more carriers on the cold end of the material. This gradient in carrier concentration leads to a potential difference, or voltage, across the material. This potential difference as a function of temperature ($\Delta V / \Delta T$) is known as the Seebeck coefficient, or thermopower.

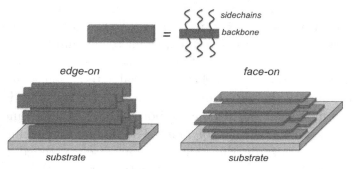

Figure 2.3. Schematic showing chain orientation in thin films of semiconducting polymers. In many semiconducting polymers, the conjugated backbone of the polymer aligns forms crystalline domains with the flat π-conjugated faces aligning. Depending both on the polymer and the subsequent processing conditions this π–π stacking direction will align in either a face-on or edge-on configuration with respect to the substrate.

direction for the best assessment of the thermoelectric performance of a material.

Deposition methods can have a significant influence on morphology and thereby charge transport. Many early studies of doped semiconducting polymers were processed using bulk methods, e.g., fiber drawing or casting bulk materials followed by stretch alignment of films with thickness $>10\,\mu$m that are less commonly used now. The dominant methods to form organic semiconducting films currently involve casting from a solution either by spin coating or by other printing methods, e.g., jet-printing, into thin films (thickness $\sim$50 nm–1 μm). There have been relatively few detailed studies to compare transport properties of materials in "bulk" relative to thin film form making it difficult to assess how transferrable results from older and more recent studies are.

2.3 Transport Models for Organic Semiconductors

Organic materials are held together by van der Waals forces making intermolecular electronic coupling relatively weak compared to covalent inorganic solids.[27] Because of this weak electronic coupling, the electronic bands of an ideal crystalline organic solid are relatively narrow ($<\sim$1.0 eV) compared to inorganic semiconductors.[28,29] The impurity levels in organic materials are generally higher than many inorganic materials due to synthetic procedures and the difficulty of purifying structurally similar organic materials.[30,31] In addition, because of the weak intermolecular forces, structural perfection is relatively rare leading to a relatively high structural defect density.[29] Band tails in single crystals of organic semiconductors have been shown to be similar to amorphous inorganic semiconductors in some cases.[32] Because most organic materials are studied in a polycrystalline form, domain or grain boundaries will also affect charge transport. Therefore, the study of the electronic structure of organic semiconductors usually relies on hopping conduction models or transport theories that have been developed for amorphous semiconductors.[28]

The apparent carrier mobilities of holes and electrons in organic solids depend on the carrier concentration due to the presence of electronic disorder and can range from 1×10^{-4} to $10\,\text{cm}^2\text{V}^{-1}\text{s}^{-1}$.[33] Transport in organic materials has been extensively examined for undoped (or very lightly doped) thin films either by inducing charge carriers via an applied gate bias in a TFT or by performing space

polaron   *bipolaron*

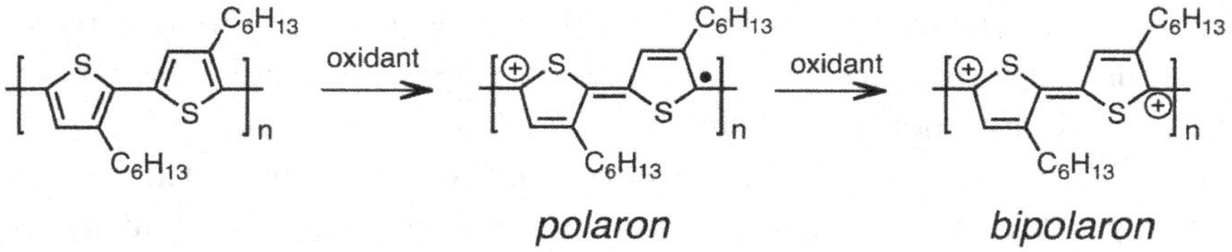

Figure 2.4. Oxidation of a P3HT leads to formation of a polaron due to structural distortion leading to stabilization of a quinoidal form. Further oxidation causes condensation of polarons into bipolarons. Bipolarons are formed in many semiconducing polymers.

charge limited current (SCLC) or time-of-flight (TOF) measurements at low carrier concentration.[28,33] These measurements have generally revealed the presence of electronic tail states with low mobility that are filled before the states with highest carrier mobility. At higher charge-carrier concentration, there is equilibrium between the polaronic and bipolaronic carriers in semiconducting polymers (Fig. 2.4) causing additional complexity in understanding the nature of charge carriers. The formation of bipolarons leads to spinless carriers, making the determination of the charge-carrier concentration difficult in many doped semiconducting polymers relative to molecular materials.

For conductive materials, the electrical transport mechanism is often inferred by the temperature dependence of the conductivity, given in a general form by.[34]

$$\sigma = \sigma_0 \exp\left[-\left(\frac{T_0}{T}\right)^{\gamma}\right]. \tag{2.1}$$

Here, T_0 is related to the activation energy, $E_A = T_0 k_B$, and γ depends on the dimensionality of transport. When $\gamma = 1$, the conductivity takes an Arrhenius form common for thermally activated transport in many organic materials. In more highly conductive samples, the exponent varies and tends to be less than 1; for example, the assumption of one-dimensional variable range hopping transport leads to $\gamma = 1/4$.

Two of the most popular models for charge transport are hopping within electronic states with Gaussian disorder[35] or a mobility edge formalism where there is an exponential tail of states away from a band edge.[33] The mobility edge model has been applied extensively to amorphous inorganic semiconductors[34] and has been successfully used to model organic TFTs formed with polymers, molecules, and single crystals. In this model, a tail of states with zero or negligible mobility exists below cut-off energy above which the states have a band-like, high mobility. In real materials, it is unlikely that there is no transport between tail states and hopping in the tails has been included in some studies. These models all share a feature of an energetic distribution of states with varying mobility, but differ in the quantitative prediction of the temperature dependence of transport. At higher carrier density, variable range hopping models have been applied based on the temperature dependence of transport.[34,36,37]

To understand the predicted behavior of these transport models for thermoelectric behavior, a general description of thermopower, α, is helpful (Eq. (2.2)).[38] Here, E_F is the Fermi energy and $\sigma(E)$ is the conductivity at a particular energy, E, that is a function of temperature and the carrier mobility, $\mu(E)$ and the density of states $g(E)$, and the Fermi function $f(E)$ (Eq. (2.3))

$$\alpha = \frac{k_B}{e} \int \frac{E - E_F}{k_B T} \frac{\sigma(E)}{\sigma} dE, \qquad (2.2)$$

$$\sigma(E) = eg(E)\mu(E)f(E)[1 - f(E)]. \qquad (2.3)$$

Using the relationship of the carrier mobility to energy for the transport models discussed, one can derive temperature-dependent expression for the thermopower. Assuming a mobility edge model for holes, the thermopower has an inverse relationship to temperature where $A > 0$ and usually between 2 and 4[34,38]:

$$\alpha = \frac{k_B}{e} \left(\frac{E_F - E_V}{k_B T} + A \right). \qquad (2.4)$$

The mobility-edge model also assumes activated transport for conductivity, such that $\gamma = 1$ in the Mott equation. In the case of hopping transport in a very narrow band, A is very small and can be assumed to be negligible, but is related to the shape of the DOS.[20] In contrast, the variable range hopping model gives a square-root dependence on T for the thermopower

$$\alpha = \frac{k_B^2}{2e}(T_0 T)^{1/2} \left. \frac{d\ln N}{dE}\right|_{E_F}, \tag{2.5}$$

where T_0 is the activation energy from the Mott equation for conductivity, with $\gamma = \frac{1}{4}$, and $d\ln N/dE|_{E_F}$ is related to the slope of the DOS at the Fermi energy. If conduction takes place near (within a few $k_B T$) the Fermi level, the temperature dependence is linear as is for metallic conduction[39]

$$\alpha = \frac{\pi^2}{3}\frac{k_B^2 T}{e} \left. \frac{d\ln\sigma}{dE}\right|_{E_F}. \tag{2.6}$$

In many cases, it is difficult to directly experimentally determine the carrier density and form of the DOS in organic semiconductors, leaving the parameters in Eqs. (2.4)–(2.6) that are related to the shape of the DOS unknown. One can use the trap-free carrier mobility in the undoped material as an upper bound for the mobility to estimate the carrier density at a given electrical conductivity. Such an assumption is likely more reasonable for polymers, where estimates of the carrier mobility in heavily doped polymers are in reasonable agreement with those from field-effect measurements, but significantly more uncertain in molecular semiconductors where the local electronic coupling is likely more sensitive. Additionally, there have been relatively few studies of the temperature dependence of the thermopower and electrical conductivity that provide information about the electronic DOS for systems other than polyacetylene and polyaniline.[36,40,41] Therefore, it is generally nontrivial to determine a theoretical relation between thermopower and the electrical conductivity.

In organic materials examined for thermoelectrics, dopants are introduced into the bulk to form carriers at high density. Because the carrier concentration is significantly higher than that studied in undoped or field-effect-induced conductivity, one expects that electronic tail states used in models of neat films would be filled. However, dopants are expected to change the structural order of the host material because of their size and shape, making it difficult to assume that the DOS determined for an intrinsic material is appropriate for a doped material. In some blends of polymers with acceptors, structurally ordered samples have been observed. For example, in iodide doped polyacetylene or poly(3-alkylthiophene) the iodine atoms will intercalate into the sidechains preserving an ordered structure.[42–45] In films of poly(3-hexylthiophene) (P3HT), the addition of F_4TCNQ results in the formation of both doped and undoped ordered domains.[46] Recently, acceptors as large as C_{60} have been demonstrated to co-crystallize with poly(alkyl-thiophene-*co*-thienothiophene) making it difficult to predict the ordering based on the molecular structure.[47]

The DOS of doped organic semiconductors is clearly perturbed by-structural disorder, making it difficult to assess the expected change in electrical behavior due to the coupling of structure and doping. To illustrate how doping affects the density of states $g(E)$, Arkhipov has developed a DOS model for hopping in the presence of charged states (counterions), with the limited assumption of a homogenous microstructure.[48] The heterogeneous nature of many organic semiconductors complicates this analysis; however, it has been used recently to develop predictions to optimize the thermo-electric figure of merit of organic semiconductors.[49]

2.4 Thermoelectric Performance of Organic Semiconductors

The thermoelectric performance of an increasing number of solution-processable organic semiconductors has been studied recently. There have been significantly more studies of p-type conduction than n-type conduction due to the historical difficulty of making materials where electron carriers (chemically reduced molecules) are stable in the ambient. Significantly more routes for doping materials to be p-type (oxidants) are available than n-type (reductants). While many organic semiconductors are now known to be ambipolar with similar hole and electron mobilities, there have not been any reports to our knowledge of the thermopower of holes and electrons in the same material beyond polyacetylene.

2.4.1 p-type materials

2.4.1.1 *Polyacetylene and polyaniline*

The thermoelectric properties of the foundational conducting polymers, polyacetylene, and polyaniline, were initially explored to understand mechanisms of charge transport as opposed to optimization for applications. Temperature-dependent conductivity and thermopower were studied for these materials extensively as a function of doping, alignment, or a combination in thin film, fibril, and bulk samples. Polyacetylene and other similar materials can be doped to very high conductivities while still maintaining a measureable thermopower. Dopants can be incorporated into the polymer film by either exposing the film to a dopant vapor such as iodine or AsF_5, or immersing the film into a solution containing a dopant ion such as $FeCl_4^-$. These methods can increase conductivity over 11 orders of magnitude from the pristine polymer, approaching, and exceeding $1,000\,Scm^{-1}$ depending on the dopant chosen, allowing calculated power factors (PFs) that are competitive for applications.[50-54] One of the highest calculated PFs reported is $1.2\,mWm^{-1}K^{-2}$ for iodine-doped stretch-aligned polyacetylene.[52] However, this method of doping requires a constant overpressure of iodine vapor, as de-doping occurs rapidly upon the removal of the dopant vapor, limiting the usefulness for applications. Percent doping, from which carrier concentration is inferred, and determined by electron spin resonance (ESR), elemental analysis, or mass increases upon doping.

Temperature-dependent transport studies of these foundational conducting polymers indicate mixed transport mechanisms.[36,50] In general, moderately and highly doped samples show a thermopower that is intermediate between sublinear ($\propto T^{1/2}$, Eq. (2.5)) and linear ($\propto T$, Eq. (2.6)) dependence on temperature, indicating primarily metallic conduction.[50] In contrast, temperature-dependent conductivity measurements indicate highly disordered transport (fluctuation induced tunneling or variable range hopping), with or without a crossover to metallic transport (and a change of sign with temperature) at high temperatures. Many of the apparent contradictions in the temperature-dependent thermopower and conductivity could be explained by different regions of order weighting differently to total thermopower and to total conductivity, which is indicative of the heterogeneous nature of these materials and other classes of semiconducting polymers.[50]

2.4.1.2 *Poly(3-hexylthiophene)*

Polythiophenes were studied originally as doped electrically conductive polymers, but have more recently been extensively studied in TFTs and OPVs.[3,6,33] Polythiophenes have higher environmental stability than polyacetylene, while still maintaining high charge-carrier mobilities. Doped poly(3-alkylthiophenes) can have electrical conductivities near $\sim 1,000\,\mathrm{Scm^{-1}}$ making them good candidates for thermoelectric applications.[55]

P3HT is the most widely used polymer in organic electronics and can be chemically doped before or after deposition into films. In many cases, solid films are soaked in solutions of a dopant to increase their electrical conductivity. The carrier concentration is usually not measured directly, but inferred from optical spectroscopy, due to the known reduction bleaching of the main optical absorption, or by the analysis of X-ray photoelectron spectroscopy (XPS), which assumes that the surface composition represents the bulk, and given as a "percent doping." A direct comparison of samples is difficult due to the problem of determining whether the difference in conductivity is a result of the apparent carrier mobility or the carrier concentration. In films of P3HT doped with $\mathrm{NO^+PF_6^-}$, the Seebeck coefficient was found to be proportional to the log of conductivity with σ ranging from 10^{-5} to $\sim 1\,\mathrm{Scm^{-1}}$, with a maximum PF of $0.1\,\mu\mathrm{Wm^{-1}K^{-2}}$ between 20% and 30% doping (Fig. 2.5).[56] In other reports, a similar logarithmic dependence was found in samples doped by Fe(III)triflimide;[57] however, there was a break in the relationship near $\sim 1\,\mathrm{Scm^{-1}}$. The maximum PF obtained was $\sim 30\,\mu\mathrm{Wm^{-1}K^{-2}}$ at $\sigma \sim 180\,\mathrm{Scm^{-1}}$. The reported thermopower was $\sim 120\,\mu\mathrm{VK^{-1}}$ for samples with $\sigma \sim 1\,\mathrm{Scm^{-1}}$, which is higher than that reported by doping with $\mathrm{NO^+PF_6^-}$, $\sim 30\,\mu\mathrm{VK^{-1}}$. Such data indicate a substantial variation in thermoelectric properties for the same

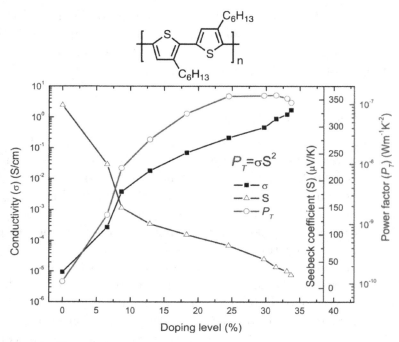

Figure 2.5. Electrical conductivity, thermopower, and power factor for P3HT as a function of percent doping by NO^+PF6^- Figure from Ref. 56, Copyright 2010 by the American Physical Society.

materials system, making it difficult to assess the limits for a given material. It is unclear if the PF maximizes at a particular carrier concentration or if the observed behavior is due to the differences in morphology.

2.4.1.3 *Co-polymers*

Most current high carrier mobility semiconducting polymers have backbone structures with co-monomers used to tune their electronic structures. There have been studies of only a few polymers relative to the broad range of materials developed for OPVs.[9] A direct comparison of materials is difficult due to different processing conditions, doping methods, and reported data. There have been several studies across materials with comparable doping methods that provide an insight into changes in thermoelectric performance in a systematic way.

Poly(2,7-carbazole)-based co-polymers have been highly successful in OPVs and have been studied as thermoelectrics using FeCl$_3$ as the dopant.[58] Poly[N-9'-heptadecanyl-2,7-carbazole-alt-5,5'-(4',7'-di-2-thie-nyl-2',1',3'-benzothiadiazole), PCDTBT and poly[N-9-heptadecanyl-2,7-carbazole-alt-5,5-(2,2-(1,4-phenylene)dithiophene)], PCDTB, were found to have similar optimal PFs, 19 μWm^{-1}K^{-2} and 14 μWm^{-1}K^{-2}, respectively. The electrical conductivity of PCDTBT was $\sim$2$\times$ higher than that of PCCDTB, but the higher thermopower of the latter led to the comparable PF. Doped films of PCDTBT were found to have good stability in the ambient with a relatively stable PF of $\sim$17 μWm^{-1}K^{-2} over 350 h despite drifts in the thermopower and electrical conductivity. While this value is lower than that required in applications, the results demonstrate favorable potential for the stability of thermoelectric polymers with high IE (deep HOMOs).

A comparison of co-polymers doped by Fe(III)triflimide has been reported where the IE of the polymers varied from 5.1 eV to 5.4 eV.[59] The Seebeck coefficient was found to decrease with the electrical conductivity for P3HT, PBTTT, PDPP3T, and PSBTBT with the largest differences between materials observed at the electrical conductivity below $\sim$1 Scm^{-1} (Fig. 2.6). The Seebeck coefficients of PBTTT and P3HT differed by a factor of $\sim$3 (13 versus 31 μVK^{-1}) at comparable electrical conductivities $\sim$200 Scm^{-1}. In a previous study with PBTTT doped by NOPF$_6$, a similar thermopower was observed at much lower electrical conductivity (54 Scm^{-1}).[60] The temperature dependence of the Seebeck coefficient was measured for the systems and found to decrease with increasing doping, suggesting a more flat DOS with more conductive samples for these materials. The maximum PFs reported ranged from $\sim$3 μWm^{-1}K^{-2} to 30 μWm^{-1}K^{-2} despite a factor of 10 difference in the highest reported electrical conductivities. There is no obvious peak in the PF for these systems as a function of electrical conductivity (carrier concentration), unlike that expected for conventional semiconductors.

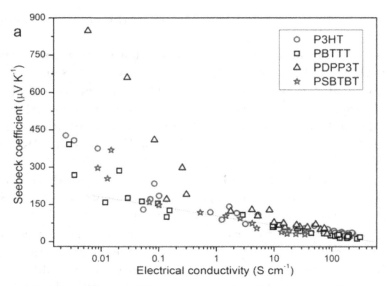

Figure 2.6. Comparison of thermopower as a function of electrical conductivity for several semiconducting polymers doped by iron triflimide. Reprinted with permission from Ref. 59. Copyright (2014) American Chemical Society.

2.4.1.4 *Poly(ethylenedioxythiopene)*

Polymers based on the ethylenedioxythiophene (EDOT) monomer, referred to as PEDOTs, have attracted much attention as thermoelectric materials.[19] There are many synthetic routes to form PEDOT-based materials for thermoelectrics. The monomer EDOT can be polymerized in the presence of an acidic polyelectrolyte, typically polystyrenesulfonic acid (PSS) to form a conductive dispersion processable from water (PEDOT:PSS). An alternative method is to polymerize EDOT with a water/alcohol solution of Fe(III)tris-*p*-toluenesulphonate leading to a form referred to as PEDOT:Tos. Prepolymerized PEDOT:PSS is commercially available and is widely used as a hole injection layer in OLEDs and OPVs. PEDOT can also be synthesized by chemical vapor deposition by exposing a film of an insulating polymer, e.g., a glycol-based triblock co-polymer, with dispersed Fe(III)Tos to EDOT vapor or by other variations.[61] Independent of the method of synthesis, PEDOT-based materials are currently the highest performing p-type organic thermoelectrics. For example, the solution-phase polymerization of PEDOT:Tos with a poly(ethyleneglycol)-based triblock co-polymer template leads to films with a conductivity of $1{,}355\,\mathrm{Scm^{-1}}$, a thermopower of $80\,\mu\mathrm{VK^{-1}}$, and a PF of $860\,\mu\mathrm{Wm^{-1}K^{-2}}$.[62]

There has been no single approach to optimizing PEDOT for thermoelectric applications, but the general goal is to increase the Seebeck coefficient while maintaining high electrical conductivity. The highest ZT (~0.4) reported yet is from aPEDOT:PSS derivative, where the total volume ratio $r\chi$ of the material is minimized to improve performance.[63] To minimize $r\chi$, the nonionized dopant species of PSS are removed from the PEDOT:PSS blend with an ethylene glycol rinse. The reduction in film thickness is attributed to the removal of PSS, which is confirmed by XPS. Interestingly, optimizing the volume ratio $r\chi$ is shown to improve all parameters associated with ZT (σ, κ, α) leading to a PF of $469\,\mu\mathrm{Wm^{-1}K^{-2}}$. Another approach used to optimize the thermoelectric properties of PEDOT has been to control its oxidation level by de-doping it chemically after synthesis. This approach has recently been used to study the thermoelectric performance of PEDOT:Tos.[61] Oxidation from 15% to 40% increases conductivity by five orders of magnitude, but decreases thermopower by two orders of magnitude (Fig. 2.7).

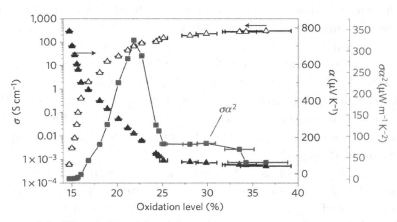

Figure 2.7. Electrical conductivity, thermopower, and power factor for PEDOT:Tos de-doped by tetrakis(dimethylamino)ethylene (TDAE). De-doping reveals a maximum in the thermoelectric power factor, $\sigma\alpha^2$ of $324\,\mu\mathrm{Wm^{-1}K^{-2}}$ at 22% oxidation. Reprinted by permission from Macmillan Publishers Ltd: Ref. 61, copyright (2011).

The power factor is maximized at 23% oxidation to a value of $350\,\mu\mathrm{Wm^{-1}K^{-2}}$. The high performance of PEDOT systems has been suggested to arise from the formation of a semimetallic state due to the interplay between polarons and bipolarons.[64]

There are many structurally similar monomers to EDOT that have been studied as electrochromics, but not examined as thermoelectrics.[19] In a recent study of poly(3,4-ethylenedioxselenophene) films formed by electrodeposition, good thermoelectric performance was found, e.g., PFs of $\sim350\,\mu\mathrm{Wm^{-1}K^{-2}}$ with the electrical conductivity of $\sim300\,\mathrm{Scm^{-1}}$.[65] These results suggest the general promise of EDOT-based systems for p-type thermoelectrics.

2.4.1.5 *Structure property relationships*

Due to the difficulty of comparing the results from different labs and exact determination of the carrier concentration, a comparison of the thermopower to the electrical conductivity can be fruitful. Such relationships have been previously discussed for inorganic semiconductors and can be related to the band structures of materials; however, here they are best considered as empirical at the current state of knowledge.[66] Individual studies of these polymers have yielded different relationships of the thermopower to the electrical conductivity at room temperature. A multisource compilation of polyacetylene data found a general trend of $\alpha \propto \sigma^{-1/4}$. In contrast, others have found an $\alpha \propto \ln\sigma$ dependence for individual doping series of relaxed and stretch-aligned polyacetylene, polyaniline, and polypyrrole.[67] It is worth noting that many of the polymers that exhibited high PFs were physically aligned samples, e.g., stretched films, which may not be advantageous due to increased thermal conductivity along the direction of transport.

Using recent data from the literature, a striking empirical relationship between the electrical conductivity and PF ($\alpha^2\sigma$) for a broad range of doped organic materials — from polymers to single crystal molecular charge transfer salts — emerges similar to initial observations on polyacetylene $\alpha \propto \sigma^{-1/4}$ (Fig. 2.8). Particularly notable is the fact that these materials exhibit transport mechanisms ranging

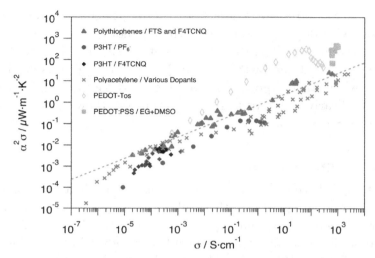

Figure 2.8. Power factor as a function of electrical conductivity for a range of semiconducting polymers. The dashed line is an empirical trend based on a power law discussed in the text. Data were taken from Refs. 17, 56, 61, 63, and unpublished work by the authors.

from hopping to near metallic conduction. It is unclear why such a relationship occurs because it is not readily predicted from existing transport models. One possibility is that the materials are single band conductors and the electronic DOS is similar, leading to the observed scaling with electrical conductivity. At this point, this relationship appears only to be defied by PEDOT-based systems, suggesting a fertile area for exploration to uncover the origin of the apparent empirical relationship.

The role of morphology on the electronic DOS in neat materials has not been perfectly defined and the study of heterogeneous systems is just beginning. In a study aimed to rationally perturb the electronic DOS, an additive P3HHT was added to P3HT to deliberately modify the DOS due to the differing IE of the two polymers. In this case, P3HHT defines the Fermi level, whereas it is assumed that the most of the current is carried in the bulk P3HT. The molecular acceptor, F_4TCNQ, was used to introduce charge carriers by electron transfer due to its high EA (low-lying LUMO).

A regime was observed where both thermopower and conductivity increase with doping modulation, which is not observed in the neat P3HT system. However, the PF peaks at $7.86\,\mathrm{nWm^{-1}K^{-2}}$ for 2 wt% P3HHT in P3HT (at 1 wt% F4TCNQ in P3HT), which is not in the regime where thermopower and conductivity are increased simultaneously. Clearly, more work is needed to uncover how rational control of the DOS can be achieved.

2.4.2   n-type materials

Traditional thermoelectric devices consist of both a hole- and an electron-conducting legs with similar thermoelectric compatibility factors to enable maximum power generation.[1] However, fewer n-type organic materials systems have been studied because the typically small EAs ($-3\,\mathrm{eV}$ to $-4\,\mathrm{eV}$) of molecular conductors make doping to carrier concentrations suitable for thermoelectric applications challenging. Historically, there are several classes of n-type molecular conductors for which the relationship between the conductivity, σ, and the Seebeck coefficient, α, has been studied.

2.4.2.1   *Charge transfer salts*

A charge transfer salt is a co-crystal of two molecules: a donor molecule such as tetrathiafulvalene (TTF) and an acceptor molecule such as tetracyanoquinodimethane (TCNQ).[68] The donor molecule will transfer the charge to the acceptor molecule depending on the difference in IE, EA, and the crystal structure. The electronic properties are highly anisotropic, and both the conductivity and thermopower vary widely along differing crystal directions (Table 2.2).[69–72] Other charge transfer salts also exhibit n-type thermoelectric behavior, in particular a wide variety of compounds based on TTF; some examples have been provided in Table 2.2. Modifying the carrier concentration of these compounds is difficult due to the generally fixed stoichiometry of co-crystals. While these materials are difficult to process directly, they can be blended with binder polymers to improve processability. For example, TTF–TCNQ blended with polyvinylchloride exhibited Seebeck coefficient of $-48\,\mu\mathrm{VK^{-1}}$ and

Table 2.2. Thermoelectric properties of selected charge transfer salt crystals at 300 K.

Material	σ (S/cm)	α (μV/K)	P.F. μWm^{-1}K^{-1}	Ref.
TTF-TCNQ				
a-axis	500	-28	40	71
b-axis		$+15$		
		-28		
TSeF-TCNQ	550	-10	6	
(BEDT-TTF)Cu$_2$Br$_4$	10^{-2}	-850	0.7	

was used as the n-type leg in an all-organic thermoelectric device with PEDOT:Tos as the p-type leg, though it has not been optimized further for practical device applications.[61]

2.4.2.2 *Fullerenes*

Fullerenes (e.g., C$_{60}$) are electron-transporting molecular semiconductors that must be extrinsically doped in order to achieve reasonable conductivities for thermoelectric applications. A variety of n-type dopants have been used to increase the carrier concentration of

Table 2.3. Structure of materials examined as n-type thermoelectrics and representative thermoelectric performance near 298 K.

Material	Dopant or counterion	σ (S/cm)	α (μV/K)	P.F. μWm^{-1}K^{-1}	Ref.
polyacetyelene (PA)	Electrochemical	≈ 5	-43.5	≈ 1	
polyaniline	H_2SO_4 vapor	6.3	-3	5.7×10^{-3}	
P(T2-NDIOD)	N-DMBI	8×10^{-3}	-850	0.6	89
Poly[K$_x$(Ni-ett)]	potassium	40	-122	60	85
C$_{60}$	Cr$_2$(hpp)$_4$ (a)	4	-175	12	
	self-doped	0.5	-170	1.4	80

[a] chromium complexed with 1,3,4,6,7,8-hexahydro-2H-pyrimido[1,2-a]pyrimidine.

fullerenes, including alkali metals,[73,74] organic dimetal complexes,[75] organometallic dimers,[76,77] and 1H-benzoimidazole derivatives.[78,79] Several dopants have been designed to be environmentally stable, notably the organometallics, while others have been designed to be compatible with solution processing. Perhaps most notably, potassium forms a distinct stoichiometric doped phase with both C_{60} and C_{70}, resulting in conductivities $\approx 500 \, \text{Scm}^{-1}$.[74] The corresponding thermoelectric properties of doped fullerenes have been studied typically for the purpose of understanding their electronic properties. Studies of the thermoelectric properties of both C_{60} and C_{70} show comparable thermoelectric PFs to the largest achieved with TTF–TCNQ, $\sim 10^{-5} \, \text{Wm}^{-1}\text{K}^{-2}$, and at the comparable electrical conductivity.[74,75]

2.4.2.3 *Perylenediimides*

Recently, a class of self-doped solution-processable molecular semiconductors based on perylenediimides (PDIs) have demonstrated n-type thermoelectric performance.[80] PDIs have high EAs $(\sim -4.0 \, \text{eV})$ and are known to have good environmental stability in an undoped form.[81] PDIs can exhibit self-doping via dehydration if the cationic amine groups are tethered to the conjugated core with a hydroxide counter ion.[82] Using this structural motif, the electrical conductivity of PDI-based materials can be tuned by changing the sidechain length resulting in an electrical conductivity (σ) of up to $0.5 \, \text{Scm}^{-1}$ and a corresponding thermoelectric PF $(\alpha^2\sigma)$ of up to $1.4 \, \mu\text{WmK}^{-2}$.

A study of structurally similar PDIs has revealed details about the connection between electrical conductivity and thermopower. While σ can be tuned over several orders of magnitude, both the Seebeck coefficient and the spin concentration detected via electron paramagnetic resonance remain relatively constant, with $\alpha \approx 200 \, \mu\text{VK}^{-1}$ (Fig. 2.9).[80] Characterization of PDI morphology with grazing-incidence wide-angle X-ray scattering reveals that tuning the side-chain length leads to dramatic changes in film morphology. It is hypothesized that these changes in morphology, for instance, in crystallite size, lead to changes in the apparent electron mobility and are responsible for dramatic changes in conductivity despite the carrier concentration remains constant. This work reveals the essential role rational molecular design and processing will play in the further development of molecular thermoelectric materials.

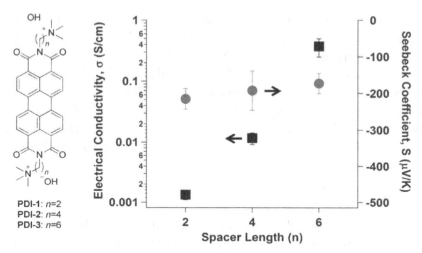

Figure 2.9. The electrical conductivity of self-doped perylene diimides (left) was found to be sensitive to the length of alkyl chains tethering charged groups to the conjugated core while the seebeck coefficient remained relatively constant, likely due to significant changes in film morphology while the carrier concentration remained relatively constant. Figure reproduced from Ref. 80 copyright 2014 WILEY-VCH Verlag GmbH & Co. KGaA, Weinheim.

2.4.2.4 *n-type polymers*

The thermoelectric properties of electron-conducting n-type semiconducting polymers have been studied significantly less than their hole-conducting p-type counterparts discussed above. This is due, in part, to the poor stability of early generation semiconducting polymer such as polyacetylene and poly(phenylenevinylenes). While an n-type thermoelectric PF can be extracted from early measurements of both polyacetylene[41] and polyaniline,[83] neither of these polymers exhibits a thermoelectric PF that exceeds that of fullerenes or charge transfer salts. Additionally, neither doping scheme is particularly stable — the polymer returns to p-type as the polymer de-dopes in the lab environment, in some cases within a matter of minutes. A variety of new electron donors have designed to enable stable extrinsic n-doping of conjugated polymers, including organometallics[77] and small molecule dopants.[84] Additionally, self-doped polymers such as the organometallic poly(K_x Ni 1,1,2,2-ethenetetrathiolate) have shown potential as thermoelectric materials; however, the low solubility necessitates solid state powder processing.[85]

Recent progress in new dopants and semiconducting polymers with high EAs (low-lying conduction levels) suggests that stable extrinsic n-type doping to carrier concentrations relevant for thermoelectric applications (10^{19}–10^{21} cm^{-3}) might be feasible, enabling the optimization of the thermoelectric figure of merit by tuning the carrier concentration. A wealth of promising new polymer backbone structures exhibit high electron mobilities in TFTs; notably, diimide-based polymers have exhibited μ of almost $1\,\mathrm{cm^2V^{-1}s^{-1}}$.[86,87] In another class of polymers based on pyromelliticdiimide, n-type TFT measurements have been corroborated by measurements of the Seebeck coefficient that confirm that electrons are the majority carrier.[88] Evaluation of n-type and ambipolar semiconducting polymers as electron-conducting thermoelectric materials is ongoing and essential to develop a viable counterpart to p-type polymers like PEDOT.

We studied a high-performance solution-processable n-type polymer, poly{N,N'-bis(2-octyl-dodecyl)-1,4,5,8-napthalenedicarboximide-2,6-diyl]-*alt*-5,5'-(2,2'-bithiophene) P(NDIOD-T2) doped with dihydro-1H-benzoimidazol-2-yl (N-DBI) derivatives.[89] At $9\,\mathrm{mol\%}$ doping, we measure $\sigma = 8 \times 10^{-3}\,\mathrm{Scm^{-1}}$ and $\alpha = -850\,\mu\mathrm{VK^{-1}}$; the sign of the thermopower is consistent with n-type conduction. We achieve a thermoelectric PF, $\alpha^2\sigma$, of $0.6\,\mu\mathrm{Wm^{-1}K^{-2}}$. To date, this is the highest n-type thermoelectric PF achieved with a solution-processable polymer, and comparable to p-type polymers at similar conductivities. However, we estimate that the maximum carrier concentration achieved in these films is $\sim 10^{17}\,\mathrm{cm^{-3}}$, indicating only 1% of all dopants added in solution contribute a mobile electron. This low carrier concentration is consistent with the relatively high measured thermopower that indicates that the Fermi level is far from the conduction level. Material characterization including atomic force microscopy, transmission electron microscopy, and grazing-incidence wide-angle X-ray scattering indicates that the dopant molecules are insoluble in the P(NDIOD-T2), forming aggregates on the surface of the film instead of homogenously dispersing within the film (Fig. 2.10). Thus, developing better miscibility between the dopant

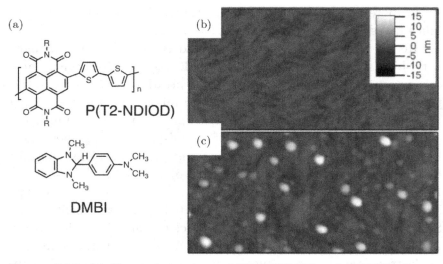

Figure 2.10. (a) Chemical structures of P(NDIOD-T2) and N-DMBI. Atomic force microscopy images of (b) neat P(NDIOD-T2) and (c) P(NDIOD-T2) doped with 9 mol% N-DMBI in solution (images are 1 μm wide and 500 nm tall). Aggregates form on the surface of the doped films, and likely contribute to a low doping efficiency that limits the electrical conductivity and thermoelectric power factor achieved. Figure reproduced from Ref. 89 copyright 2014 WILEY-VCH Verlag GmbH & Co. KGaA, Weinheim.

and polymer-host will likely be critical to achieve carrier concentrations $\sim 10^{19}$–10^{21} cm^{-3} in n-type polymeric thermoelectric materials.

2.5 Thermal Conductivity of Organic Semiconductors

The thermal conductivity, κ, of a material can be separated into the heat carried by phonons of the lattice, κ_l, and by electrons, κ_{elec}. The overall thermal conductivity is a sum of these factors because they operate in parallel. Additionally, the thermal and electrical conductivities of a material are tensors and may not have the same relative magnitude in the same direction. Anisotropies are particularly important for organic materials where the molecules in most samples are well oriented due to the texture of crystalline domains relative to the substrate and to chain alignment in polymers. However, many thermal measurement techniques, such as 3ω and time decay thermal reflectance (TDTR), only provide the cross-plane thermal conductivity, whereas the electrical properties are usually measured through plane.[90,91] The molecular anisotropy, therefore, makes determination of ZT problematic in films of organic semiconductors.

The thermal conductivity, κ_l, of both random coil and stiff chain insulating polymers has been studied. The thermal conductivity of amorphous electrically insulating polymers is typically $\sim 0.3\,\mathrm{Wm^{-1}K^{-1}}$ near room temperature.[92,93] Aligned, ordered polymers are known to have higher thermal conductivities. For instance, measurements of spin-coated polyimide, which is structurally similar to many semiconducting polymers and has similar chain alignment relative to the substrate, give the cross-plane thermal conductivity of $\sim 0.3\,\mathrm{Wm^{-1}K^{-1}}$, whereas the in-plane value was between $1\,\mathrm{Wm^{-1}K^{-1}}$ and $2\,\mathrm{Wm^{-1}K^{-1}}$ depending on the thickness.[94] Recently, TDTR measurements of κ_l of high modulus aligned polymer fibers have been reported with values ranging from $2\,\mathrm{Wm^{-1}K^{-1}}$ to $20\,\mathrm{Wm^{-1}K^{-1}}$.[95] These values are significantly higher than those of amorphous coil polymers. Studies of undoped P3HT films have yielded cross-plane values ~ 0.2–$0.3\,\mathrm{Wm^{-1}K^{-1}}$ similar to those reported for polyimide and coil polymers in the bulk,[96] but there have not been reports of the through-plane thermal conductivity.

The limits of the thermal conductivity of organic semiconductors and the relationship to the electrical conductivity have not been fully examined. If one makes a very basic assumption that κ_{elec} follows a Weidemann–Franz-like law for an electron gas, electrons are only likely to make a substantial contribution to the overall thermal conductivity at electrical conductivities above $100\,\mathrm{Scm^{-1}}$.[27] For materials with very low thermal conductivity, such as solution-processed fullerenes, κ_l, $\sim 0.06\,\mathrm{Wm^{-1}K^{-1}}$,[97] the electronic component may be observable at slightly lower electrical conductivity. Therefore, we expect that κ_{elec} only contributes in samples where the electrically conductivity is high and that κ_l can be used as a good approximation below these values. Much stronger changes are likely expected based on the anisotropies observed in the absence of electrical conduction.

There are few reports of the thermal conductivity of doped semiconducting polymers at sufficiently high electrical conductivity that the thermal transport by electrons could be significant. In one study of *trans*-polyacetyelene, the undoped thermal conductivity was found to be $0.38\,\mathrm{Wm^{-1}K^{-1}}$, whereas in a doped sample (σ of $270\,\mathrm{Scm^{-1}}$), κ was $0.69\,\mathrm{Wm^{-1}K^{-1}}$.[98] These values are both lower than those observed in insulating samples of polyimide and in high modulus fibers, and it is unclear if the difference in κ is due to processing methods or the electrical contribution to the thermal conductivity. The cross-plane thermal conductivity of thin films of PEDOT:PSS, the current best thermoelectric material, has been reported to be $\sim 0.25\,\mathrm{Wm^{-1}K^{-1}}$ by TDTR.[96] Estimates of the anisotropy have been made showing relatively small differences for PEDOT:Tos and PEDOT:PSS even at high electrical conductivity ($\sigma \sim 1{,}000\,\mathrm{Scm^{-1}}$), e.g., cross-plane values of $0.33\,\mathrm{Wm^{-1}K^{-1}}$ and in-plane values of $0.37\,\mathrm{Wm^{-1}K^{-1}}$.[99,100] However, a systematic study of the thermal conductivity as a function of processing methods and electrical conductivity is still lacking. Clearly, the determination of how thermal and electronic conductivities are related in organic semiconductors is a critical area of future research.

2.6 Conclusion

Organic semiconductors are emerging as a potential alternative to inorganic thermoelectrics in applications near room temperature. Current work on the PEDOT-based system suggests strongly that organic thermoelectrics can have similar performance to inorganic materials (ZT ~ 0.4).[62,64,99] Understanding of the connections between molecular structure, micro/nanostructure, and charge/heat transport is at an early stage. Improvements in methods to stably dope organic semiconductors over a wide range of carrier concentrations will provide a means to understand charge transport mechanisms and how the electronic DOS controls the thermopower. A wide range of semiconducting polymers and molecules that have been synthesized for applications in TFTs and solar cells are opening up a wide space for the exploration of new thermoelectrics. Many of these materials have relatively high carrier mobility for both holes and electrons,[6] suggesting that it should be possible to form both legs of a thermoelectric device with the same material.

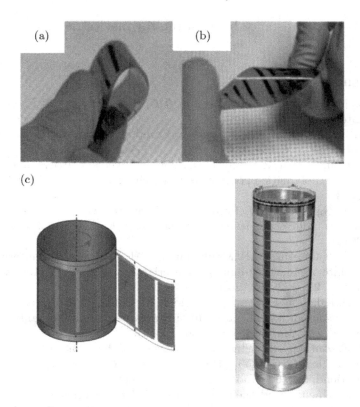

Figure 2.11. Images of flexible prototype thermoelectric modules based on PEDOT showing resistance to (a) bending and (b) torsion. Reproduced from Ref. 62 with permission of The Royal Society of Chemistry. (c) Roll-to-roll printed PEDOT:PSS thermoelectric module.

The emergence of proof-of-concept thermoelectric modules using printing methods further demonstrates the potential for the fabrication of thermoelectric waste heat scavengers and heating/cooling elements (Fig. 2.11).[61,62,101] Because these organic materials can be made deposited easily near room temperature, it is possible to imagine high throughput fabrication using method as roll-to-roll printing.[101] These early printed modules have relatively low power output, e.g., $20\,\mathrm{nWcm^{-2}}$ with a leg packing density of 0.47.[61] Such prototypes point out future areas of discovery, particularly the need for high performance n-type legs and an understanding of the role of the contact resistance between the thermoelectric and conductor linking the legs.

References

1. D. M. Rowe, *CRC Handbook of Thermoelectrics*, CRC Press: Boca Raton, FL (1995).
2. G. J. Snyder and E. S. Toberer, Complex thermoelectric materials. *Nat. Mater.*, **7**, 105–114 (2008).
3. L. Dou, J. Von, Z. hong, *et al.*, 25th anniversary article: adecade of organic/polymeric photovoltaic research. *Adv. Mater.*, **25**, 6642–6671 (2013).
4. T. N. Ng and D. E. Schwarz, Scalable printed electronics: An organic decoder addressing ferroelectric non-volatile memory. *Sci. Rep.*, **2**, (2012).
5. K. T. Kamtekar, A. P. Monkman, and M. R. Bryce, Recent advances in white organic light-emitting materials and devices (WOLEDs). *Adv. Mater.*, **22**, 572–582 (2010).
6. S. Holliday, J. E. Donaghey, and I. McCulloch, Advances in charge carrier mobilities of semiconducting polymers used in organic transistors. *Chem. Mater.*, **26**, 647–663 (2014).
7. J. E. Anthony, Functionalized acenes and heteroacenes for organic electronics. *Chem. Rev.*, **106**, 5028–5048 (2006).
8. M. Bendikov, F. Wudl, and D. F. Perepichka, Tetrathiafulvalenes, oligoacenenes, and their buckminsterfullerene derivatives: The brick and mortar of organic electronics. *Chem. Rev.*, **104**, 4891–4946 (2004).
9. H. J. Son, F. He, B. Carsten, and L. Yu, Are we there yet? Design of better conjugated polymers for polymer solar cells. *J. Mater. Chem.*, **21**, 18934 (2011).
10. A. Salleo, R. J. Kline, D. M. DeLongchamp, and M. L. Chabinyc, Microstructural characterization and charge transport in thin films of conjugated polymers. *Adv. Mater.*, **22**, 3812–3838 (2010).
11. D. M. DeLongchamp, R. J. Kline, D. A. Fischer, L. J. Richter, and M. F. Toney, Molecular characterization of organic electronic films. *Adv. Mater.*, **23**, 319–337 (2011).
12. O. Bubnova and X. Crispin, Towards polymer-based organic thermoelectric generators. *Energy Environ. Sci.*, **5**, 9345–9362 (2012).
13. P. M. Chaikin and G. Beni, Thermopower in the correlated hopping regime. *Phys. Rev. B*, **13**, 647–651 (1976).
14. P. M. Chaikin, R. L. Greene, S. Etemad, and E. Engler, Thermopower of an isostructural series of organic conductors. *Phys. Rev. B*, **13**, 1627–1632 (1976).

15. J. B. Torrance, The difference between metallic and insulating salts of tetracyanoquinodimethone (TCNQ): How to design an organic metal. *Acc. Chem. Res.*, **12**, 79–86 (1979).
16. A. J. Heeger, Semiconducting and metallic polymers: The fourth generation of polymeric materials. *J. Phys. Chem. B*, **105**, 8475–8491 (2001).
17. A. Shakouri and S. Li, Thermoelectric power factor for electrically conductive polymers, in *Eighteenth Int. Conf. Thermoelectr.*, pp. 402–406 (1999). doi:10.1109/ICT.1999.843415.
18. T. Schimmel, *et al.*, High-θ polyacetylene: DC conductivity between 14 mK and 300 K. *Synth. Met.*, **28**, D11–D18 (1989).
19. L. Groenendaal, F. Jonas, D. Freitag, H. Pielartzik, and J. R. Reynolds, Poly(3,4-ethylenedioxythiophene) and its derivatives: Past, present, and future. *Adv. Mater.*, **12**, 481–494 (2000).
20. K. Walzer, B. Maennig, M. Pfeiffer, and K. Leo, Highly efficient organic devices based on electrically doped transport layers. *Chem. Rev.*, **107**, 1233–1271 (2007).
21. J.-L. Bredas, Mind the gap! *Mater. Horiz.*, **1**, 17 (2014).
22. A. O. Patil, A. J. Heeger, and F. Wudl, Optical properties of conducting polymers. *Chem. Rev.*, **88**, 183–200 (1988).
23. A. A. Virkar, S. Mannsfeld, Z. Bao, and N. Stingelin, Organic semiconductor growth and morphology considerations for organic thin-film transistors. *Adv. Mater.*, **22**, 3857–3875 (2010).
24. J. Rivnay, S. C. B. Mannsfeld, C. E. Miller, A. Salleo, and M. F. Toney, Quantitative determination of organic semiconductor microstructure from the molecular to device scale. *Chem. Rev.*, **112**, 5488–5519 (2012).
25. J. Rivnay, *et al.*, Large modulation of carrier transport by grain-boundary molecular packing and microstructure in organic thin films. *Nat. Mater.*, **8**, 952–958 (2009).
26. A. Von Mühlenen, N. Errien, M. Schaer, M.-N. Bussac, and L. Zuppiroli, Thermopower measurements on pentacene transistors. *Phys. Rev. B*, **75**, 115338 (2007).
27. N. W. Ashcroft, *Solid State Physics*, Holt, Rinehart and Winston: Austin, TX, 1976.
28. V. Coropceanu, J. Cornil, and D. A. da Silva Filloet, Charge transport in organic semiconductors. *Chem. Rev.*, **107**, 926–952 (2007).
29. M. Pope, *Electronic Processes in Organic Crystals and Polymers*, Oxford University Press: Oxford, 1999.
30. J. E. Anthony, M. Heeney, and B. S. Ong, Synthetic aspects of organic semiconductors. *MRS Bull.*, **33**, 698–705 (2008).

31. H. E. Katz, Z. Bao, and S. L. Gilat, Synthetic chemistry for ultra-pure, processable, and high-mobility organic transistor semiconductors. *Acc. Chem. Res.*, **34**, 359–369 (2001).

32. D. V. Lang, X. Chi, T. Siegrist, A. M. Sergent, and A. P. Ramirez, Amorphouslike density of gap states in single-crystal pentacene. *Phys. Rev. Lett.*, **93**, 086802 (2004).

33. H. Sirringhaus, 25th anniversary article: Organic field-effect transistors: the path beyond amorphous silicon. *Adv. Mater.*, **26**, 1319–1335 (2014).

34. N. F. Mott and E. A. Davis, *Electronic Processes in Non-crystalline Materials*, Oxford University Press: Oxford, 1971.

35. R. Coehoorn, W. F. Pasveer, P. A. Bobbert, and M. A. J. Michels, Charge-carrier concentration dependence of the hopping mobility in organic materials with Gaussian disorder. *Phys. Rev. B*, **72**, 155206 (2005).

36. Z. H. Wang, E. M. Scherr, A. G. MacDiarmid, and A. J. Epstein, Transport and EPR studies of polyaniline: A quasi-one-dimensional conductor with three-dimensional 'metallic' states. *Phys. Rev. B*, **45**, 4190–4202 (1992).

37. S. Wang, M. Ha, M. Manno, C. Daniel Frisbie, and C. Leighton, Hopping transport and the Hall effect near the insulator–metal transition in electrochemically gated poly(3-hexylthiophene) transistors. *Nat. Commun.*, **3**, 1210 (2012).

38. H. Fritzsche, A general expression for the thermoelectric power. *Solid State Commun.*, **9**, 1813–1815 (1971).

39. A. B. Kaiser, Thermoelectric power and conductivity of heterogeneous conducting polymers. *Phys. Rev. B*, **40**, 2806–2813 (1989).

40. D. Moses, A. Denenstein, J. Chen *et al.*, Effect of nonuniform doping on electrical transport in trans-$(CH)x$: Studies of the semiconductor–metal transition. *Phys. Rev. B*, **25**, 7652–7660 (1982).

41. C. O. Yoon, M. Reghu, D. Moses *et al.*, Hopping transport in doped conducting polymers in the insulating regime near the metal–insulator boundary: Polypyrrole, polyaniline and polyalkylthiophenes. *Synth. Met.*, **75**, 229–239 (1995).

42. M. Winokur, Y. B. Moon, A. J. Heeger *et al.*, X-ray scattering from sodium-doped polyacetylene: Incommensurate-commensurate and order–disorder transformations. *Phys. Rev. Lett.*, **58**, 2329–2332 (1987).

43. S. L. Hsu, A. J. Signorelli, G. P. Pez, and R. H. Baughman, Highly conducting iodine derivatives of polyacetylene: Raman, XPS and x-ray diffraction studies. *J. Chem. Phys.*, **69**, 106–111 (1978).

44. N. S. Murthy, G. G. Miller, and R. H. Baughman, Structure of polyacetylene–iodine complexes. *J. Chem. Phys.*, **89**, 2523–2530 (1988).
45. M. J. Winokur, P. Wamsley, J. Moulton, P. Smith, and A. J. Heeger, Structural evolution in iodine-doped poly(3-alkylthiophenes). *Macromolecules*, **24**, 3812–3815 (1991).
46. D. T. Duong, C. Wang, E. Antono, M. F. Toney, and A. Salleo, The chemical and structural origin of efficient p-type doping in P3HT. *Org. Electron.*, **14**, 1330–1336 (2013).
47. N. C. Miller, E. Cho, M. J. N. Junk *et al.* Use of X-ray diffraction, molecular simulations, and spectroscopy to determine the molecular packing in a polymer-fullerene bimolecular crystal. *Adv. Mater.*, **24**, 6071–6079 (2012).
48. V. Arkhipov, P. Heremans, E. Emelianova, and H. Bässler, Effect of doping on the density-of-states distribution and carrier hopping in disordered organic semiconductors. *Phys. Rev. B*, **71** (2005).
49. G. Kim and K. P. Pipe, Thermoelectric model to characterize carrier transport in organic semiconductors. *Phys. Rev. B*, **86**, 085208 (2012).
50. A. B. Kaiser, Thermoelectric power and conductivity of heterogeneous conducting polymers. *Phys. Rev. B*, **40**, 2806–2813 (1989).
51. N. T. Kemp, A. Kaiser, C. J. Liu, *et al.* Thermoelectric power and conductivity of different types of polypyrrole. *J. Polym. Sci. Part B Polym. Phys.*, **37**, 953–960 (1999).
52. Y. Nogami, H. Kaneko, T. Ishiguro, *et al.* On the metallic states in highly conducting iodine-doped polyacetylene. *Solid State Commun.*, **76**, 583–586 (1990).
53. J. F. Kwak, T. C. Clarke, R. L. Greene, and G. B. Street, Transport properties of heavily AsF5 doped polyacetylene. *Solid State Commun.*, **31**, 355–358 (1979).
54. Y. W. Park, W. K. Han, C. H. Choi, and H. Shirakawa, Metallic nature of heavily doped polyacetylene derivatives: Thermopower. *Phys. Rev. B*, **30**, 5847–5851 (1984).
55. R. D. McCullough, The chemistry of conducting polythiophenes. *Adv. Mater.*, **10**, 93–116 (1998).
56. Y. Xuan, X. Lin, S. Desbief, *et al.* Thermoelectric properties of conducting polymers: The case of poly(3-hexylthiophene). *Phys. Rev. B*, **82**, 115454 (2010).
57. Q. Zhang, Y. Sun, W. Xu, and D. Zhu, Thermoelectric energy from flexible P3HT films doped with a ferric salt of triflimide anions. *Energy Environ. Sci.*, **5**, 9639–9644 (2012).
58. R. B. Aïch, N. Blouin, A. Bouchard, and M. Leclerc, Electrical and thermoelectric properties of poly(2,7-carbazole) derivatives. *Chem. Mater.*, **21**, 751–757 (2009).

59. Q. Zhang, Y. Sun, W. Xu, and D. Zhu, What to expect from conducting polymers on the playground of thermoelectricity: Lessons learned from four high-mobility polymeric semiconductors. *Macromolecules*, **47**, 609–615 (2014).

60. Q. Zhang, Y. Sun, F. Jiao, *et al.* Effects of structural order in the pristine state on the thermoelectric power-factor of doped PBTTT films. *Synth. Met.*, **162**, 788–793 (2012).

61. O. Bubnova, Z. U. Khan, A. Malti *et al.* Optimization of the thermoelectric figure of merit in the conducting polymer poly(3,4-ethylenedioxythiophene). *Nat. Mater.*, **10**, 429–433 (2011).

62. T. Park, C. Park, B. Kim, H. Shin, and E. Kim, Flexible PEDOT electrodes with large thermoelectric power factors to generate electricity by the touch of fingertips. *Energy Environ. Sci.*, **6**, 788–792 (2013).

63. G.-H. Kim, L. Shao, K. Zhang, and K. P. Pipe, Engineered doping of organic semiconductors for enhanced thermoelectric efficiency. *Nat. Mater.*, **12**, 719–723 (2013).

64. O. Bubnova, Z. Y. Jhan, H. Wang, *et al.* Semi-metallic polymers. *Nat. Mater.*, **13**, 190–194 (2013).

65. B. Kim, H. Shin, T. Park, H. Lim, and E. Kim, NIR-sensitive poly(3,4-ethylenedioxyselenophene) derivatives for transparent photo-thermoelectric converters. *Adv. Mater.*, **25**, 5483–5489 (2013).

66. D. M. Rowe and G. Min, α-in σ plot as a thermoelectric material performance indicator. *J. Mater. Sci. Lett.*, **14**, 617–619 (1995).

67. N. Mateeva, H. Niculescu, J. Schlenoff, and L. R. Testardi, Correlation of Seebeck coefficient and electric conductivity in polyaniline and polypyrrole. *J. Appl. Phys.*, **83**, 3111 (1998).

68. A. F. Garito and A. J. Heeger, Design and synthesis of organic metals. *Acc. Chem. Res.*, **7**, 232–240 (1974).

69. D. E. Schafer, F. Wudl, G. A. Thomas, J. P. Ferraris, and D. O. Cowan, Apparent giant conductivity peaks in an anisotropic medium: TTF-TCNQ. *Solid State Commun.*, **14**, 347–351 (1974).

70. M. J. Cohen, L. B. Coleman, A. F. Garito, and A. J. Heeger, Electrical conductivity of tetrathiofulvalinium tetracyanoquinodimethan (TTF)(TCNQ). *Phys. Rev. B*, **10**, 1298–1307 (1974).

71. J. F. Kwak, P. M. Chaikin, A. A. Russel, A. F. Garito, and A. J. Heeger, Anisotropic thermoelectric power of TTF-TCNQ. *Solid State Commun.*, **16**, 729–732 (1975).

72. G. A. Thomas, D. E. Shafer, F. Wudl, *et al.* Electrical conductivity of tetrathiafulvalenium-tetracyanoquinodimethanide (TTF-TCNQ). *Phys. Rev. B*, **13**, 5105–5110 (1976).

73. Z. H. Wang, K. Ichimura, M. S. Dresselhaus, *et al.* Electronic transport properties of K_xC_{70} thin films. *Phys. Rev. B*, **48**, 10657–10660 (1993).

74. Z. H. Wang, A. W. P. Fung, G. Dresselhaus, *et al.* Electron–electron interactions and superconducting fluctuations in weakly localized K_3C_{60}. *Phys. Rev. B*, **47**, 15354–15357 (1993).
75. T. Menke, D. Ray, J. Meiss, K. Leo, and M. Riede, In-situ conductivity and Seebeck measurements of highly efficient n-dopants in fullerene C60. *Appl. Phys. Lett.*, **100**, 093304 (2012).
76. S. Guo, S. B. Kim, S. K. Mohapatra, *et al.* n-Doping of organic electronic materials using air-stable organometallics. *Adv. Mater.*, **24**, 699–703 (2012).
77. Y. Qi, S. K. Mohapatra, S. Kim., *et al.* Solution doping of organic semiconductors using air-stable n-dopants. *Appl. Phys. Lett.*, **100**, 083305 (2012).
78. P. Wei, T. Menke, B. D. Naab *et al.* 2-(2-Methoxyphenyl)-1,3-dimethyl-1H-benzoimidazol-3-ium iodide as a new air-stable n-type dopant for vacuum-processed organic semiconductor thin films. *J. Am. Chem. Soc.*, **134**, 3999–4002 (2012).
79. P. Wei, J. H. Oh, G. Dong, and Z. Bao, Use of a 1H-benzoimidazole derivative as an n-type dopant and to enable air-stable solution-processed n-channel organic thin-film transistors. *J. Am. Chem. Soc.*, **132**, 8852–8853 (2010).
80. B. Russ, M. J. Robb, F. Brunettii, *et al.* Power factor enhancement in solution-processed organic n-type thermoelectrics through molecular design. *Adv. Mater.*, **26**, 3473-3477 (2014).
81. X. Zhan, A. Facchetti, S. Barlow, *et al.* Rylene and related diimides for organic electronics. *Adv. Mater.*, **23**, 268–284 (2011).
82. T. H. Reilly, A. W. Hains, H.-Y. Chen, and B. A. Gregg, A self-doping, O_2-stable, n-type interfacial layer for organic electronics. *Adv. Energy Mater.*, **2**, 455–460 (2012).
83. C. O. Yoon, M. Reghu, D. Moses, A. J. Heeger, and Y. Cao, Counterion-induced processibility of polyaniline: Thermoelectric power. *Phys. Rev. B*, **48**, 14080–14084 (1993).
84. N. Cho, H.-L. Yip, S. K. Hau, *et al.* n-Doping of thermally polymerizable fullerenes as an electron transporting layer for inverted polymer solar cells. *J. Mater. Chem.*, **21**, 6956 (2011).
85. Y. Sun, *et al.* Organic thermoelectric materials and devices based on p- and n-type poly(metal 1,1,2,2-ethenetetrathiolate)s. *Adv. Mater.*, **24**, 932–937 (2012).
86. J. E. Anthony, A. Facchetti, M. Heeney, S. R. Marder, and X. Zhan, n-Type organic semiconductors in organic electronics. *Adv. Mater.*, **22**, 3876–3892 (2010).
87. B. J. Jung, N. J. Tremblay, M.-L. Yeh, and H. E. Katz, Molecular design and synthetic approaches to electron-transporting organic transistor semiconductors[‡]. *Chem. Mater.*, **23**, 568–582 (2011).

88. S. Kola, J. H. Kim, R. Ireland, *et al.* Pyromellitic diimide–ethynylene-based homopolymer film as an N-channel organic fielde effect transistor semiconductor. *ACS Macro Lett.*, **2**, 664–669 (2013).

89. R. A. Schlitz, F. G. Brunetti, *et al.* Solubility-limited extrinsic n-type doping of a high electron mobility polymer for thermoelectric applications. *Adv. Mater.*, **26**, 1–6 (2014).

90. D. G. Cahill, W. K. Ford, K. E. Goodson, *et al.* Nanoscale thermal transport. *J. Appl. Phys.*, **93**, 793–818 (2003).

91. Y. K. Koh, S. L. Singer, W. Kim, *et al.* Comparison of the 3ω method and time-domain thermoreflectance for measurements of the cross-plane thermal conductivity of epitaxial semiconductors. *J. Appl. Phys.*, **105**, 054303 (2009).

92. D. G. Cahill and R. O. Pohl, Thermal conductivity of amorphous solids above the plateau. *Phys. Rev. B*, **35**, 4067–4073 (1987).

93. M. D. Losego, L. Moh, K. A. Arpin, D. G. Cahill, and P. V. Braun, Interfacial thermal conductance in spun-cast polymer films and polymer brushes. *Appl. Phys. Lett.*, **97**, 011908 (2010).

94. K. Kurabayashi, M. Asheghi, M. Touzelbaev, and K. E. Goodson, Measurement of the thermal conductivity anisotropy in polyimide films. *J. Microelectromech. Syst.*, **8**, 180–191 (1999).

95. X. Wang, V. Ho, R. A. Segalman, and D. G. Cahill, Thermal conductivity of high-modulus polymer fibers. *Macromolecules*, **46**, 4937–4943 (2013).

96. J. C. Duda, P. E. Hopkins, Y. Shen, and M. C. Gupta, Thermal transport in organic semiconducting polymers. *Appl. Phys. Lett.*, **102**, 251912 (2013).

97. X. Wang, C. D. Liman, N. D. Treat, M. L. Chabinyc, and D. G. Cahill, Ultralow thermal conductivity of fullerene derivatives. *Phys. Rev. B*, **88**, 075310 (2013).

98. D. Moses and A. Denenstein, Experimental determination of the thermal conductivity of a conducting polymer: Pure and heavily doped polyacetylene. *Phys. Rev. B*, **30**, 2090–2097 (1984).

99. G.-H. Kim, L. Shao, K. Zhang, and K. P. Pipe, Engineered doping of organic semiconductors for enhanced thermoelectric efficiency. *Nat. Mater.*, **12**, 719–723 (2013).

100. O. Bubnova, *et al.* Optimization of the thermoelectric figure of merit in the conducting polymer poly(3,4-ethylenedioxythiophene). *Nat. Mater.*, **10**, 429–433 (2011).

101. R. R. Søndergaard, M. Hösel, N. Espinosa, M. Jørgensen, and F. C. Krebs, Practical evaluation of organic polymer thermoelectrics by large-area R2R processing on flexible substrates. *Energy Sci. Eng.*, **1**, 81–88 (2013).

Chapter 3

Nanostructured Thermoelectric Materials

Sangyeop Lee and Gang Chen

The introduction of nanostructures in thermoelectric materials has led to significant improvements of the thermoelectric figure of merit, ZT, over the past two decades. Initial efforts have focused on using low-dimensional materials to engineer electron band structures and reduce phonon mean free paths. Subsequent efforts have been made to develop bulk thermoelectric materials maintaining the features of the low-dimensional materials, and the effort has led to the high ZT of bulk nanostructured materials recently reported. In this chapter, we summarize the theoretical background and the experimental demonstrations that resulted in these advancements. We also address several unanswered issues regarding electron/phonon transport in thermoelectric materials, which may hold the key to further advances in ZT.

3.1 Introduction

Thermoelectric devices have many advantages over conventional thermo-mechanical energy conversion devices due to their solid-state characteristics, such as no moving parts, high reliability, and easy scalability. These advantages have led to some noteworthy applications of of these devices, including in automotive climate control seats,[1] cooling diode lasers, potential applications in power generation from solar irradiation,[2] and waste heat recovery.[1] However, the low efficiency compared to conventional thermomechanical cycles has limited thermoelectric devices to niche applications where the conventional cycles cannot be easily applied.[3]

The maximum efficiency of thermoelectric energy conversion devices is determined by the thermoelectric figure of merit,

$$ZT = \frac{S^2 \sigma}{\kappa} T, \qquad (3.1)$$

where S, σ, κ, and T are the Seebeck coefficient, electrical conductivity, thermal conductivity, and temperature, respectively. The numerator, $S^2\sigma$, is called the thermoelectric power factor. Expression (3.1) shows three important requirements for a material to exhibit high ZT: a large thermoelectric effect, small ohmic losses, and small heat leakage through the thermoelectric material. However, improving ZT is not straightforward since the three important properties (S, σ, and κ) are interdependent. For example, small doping concentration is desirable to achieve a large Seebeck coefficient and low electronic thermal conductivity (κ_e), while it leads to low electrical conductivity. Conversely, large doping concentration is advantageous for high electrical conductivity, but decreases the Seebeck coefficient and increases the electronic thermal conductivity.

The interdependence of these three transport properties has hindered the development of efficient thermoelectric materials. Since the conventional thermoelectric materials, Bi_2Te_3 and $PbTe$, were developed in 1950s, there have been many efforts to increase ZT further. Alloy systems and carrier concentration optimization have been investigated as approaches improving the conventional thermoelectric materials, and searching for new thermoelectric materials has continued. However, since ZT ~ 1 was achieved with the alloy of the conventional thermoelectric materials in 1960s, there was no report for the noticeable improvement of ZT until 1990s.[4]

The idea of using low-dimensional thermoelectric materials, suggested by Hicks and Dresselhaus in 1993,[5,6] changed the paradigm of thermoelectric material development. They predicted the improvement of ZT by several times if the conventional thermoelectric materials were used in one-(1D) or two-dimensional (2D) structures. These predictions opened a new opportunity to engineer electron and phonon transport in thermoelectric materials beyond the past approaches largely limited to carrier concentration optimization and searching for new thermoelectric materials. In the subsequent studies, various nanostructures, such as nanowires and nanocomposites, have been applied to the conventional thermoelectric materials, and a peak value of ZT ~ 2.2 was recently reported.[7]

In this chapter, we briefly introduce both theoretical backgrounds and experimental demonstrations for high-ZT thermoelectric materials through the use of nanostructures. In the following sections, we discuss the enhancement of electronic properties in inorganic and polymer thermoelectric materials(Section 3.2) and the reduction of lattice thermal conductivity (Section 3.3). We not only focus on the benefits of nanostructures, but also introduce several successful strategies in bulk materials that possibly give valuable insights for the further improvement of ZT.

3.2 Electron Transport

In the absence of the magnetic field, electron transport is driven by the gradients of temperature and electromotive force. The thermoelectric effect is due to the coupling between these two driving forces. When the driving forces are small, the linear response theory can be used to describe electron transport as[8]

$$J_e = L_{11}\left(-\frac{d\Phi}{dx}\right) + L_{12}\left(-\frac{dT}{dx}\right), \tag{3.2}$$

$$J_Q = L_{21}\left(-\frac{d\Phi}{dx}\right) + L_{22}\left(-\frac{dT}{dx}\right), \tag{3.3}$$

where J_e and J_Q are the charge flux and heat flux, respectively. Here, Φ and T are the electromotive force (the electrochemical potential divided by the unit charge) and temperature, respectively. Based on the Boltzmann equation under the relaxation time approximation and isotropic transport, the transport coefficients tensor, L_{ij}, can be expressed with electron band structure parameters and scattering rates for crystalline materials:

$$L_{11} = \mathcal{L}^{(0)}, \tag{3.4}$$

$$L_{21} = TL_{12} = -\frac{1}{e}\mathcal{L}^{(1)}, \tag{3.5}$$

$$L_{22} = \frac{1}{e^2T}\mathcal{L}^{(2)}, \tag{3.6}$$

$$\mathcal{L}^{(\alpha)} = -\frac{e^2}{3}\int (E - E_F)^\alpha v^2\tau\frac{\partial f_0}{\partial E}D(E)dE, \tag{3.7}$$

where E_F and E are the electrochemical potential and electron energy, respectively. Here, v and $D(E)$ are the group velocity and density of states, respectively. Other variables, such as $f\tau$, and e, are the Fermi–Dirac distribution, relaxation time, and electronic charge, respectively.

The electron transport properties, such as the Seebeck coefficient and electrical conductivity, can be expressed using the transport coefficient tensor, L_{ij}. First,the electrical conductivity is defined by the relationship between the charge flux (J_e) and electromotive force gradients $(-d\Phi/dx)$ when there is no temperature gradient:

$$\sigma = \frac{J_e}{-\frac{d\Phi}{dx}} = L_{11} = -\frac{e^2}{3}\int v^2\tau\frac{\partial f_0}{\partial E}D(E)dE. \tag{3.8}$$

Under open circuit conditions, the gradients of the electromotive force $(-d\Phi/dx)$ and temperature gradient $(-dT/dx)$ define the Seebeck coefficient as

$$S = \frac{L_{12}}{L_{11}} = -\frac{1}{e}\frac{\int \left(\frac{E-E_F}{T}\right)v^2\tau\frac{\partial f_0}{\partial E}D(E)dE}{\int v^2\tau\frac{\partial f_0}{\partial E}D(E)dE}. \tag{3.9}$$

Similarly, the electronic thermal conductivity is defined by the heat flux (J_Q) and temperature gradient $(-dT/dx)$ when there is no charge flux:

$$\kappa_e = L_{22} - \frac{L_{12}L_{21}}{L_{11}} = \frac{1}{e^2T}\left(\mathcal{L}^{(2)} - \frac{\mathcal{L}^{(1)}\mathcal{L}^{(1)}}{\mathcal{L}^{(0)}}\right). \tag{3.10}$$

From Eqs. (3.8)–(3.10), we clearly see that the transport properties are the functions of (i) the electrochemical potential that determines the Fermi window $(-\partial f_0/\partial E)$ and carrier concentration, (ii) the electron band structure parameter $(v^2 D(E))$, and (iii) the relaxation time (τ). Therefore, these three parameters can be controlled to achieve the high thermoelectric power factor, and in the following sections, we will summarize how these three parameters were controlled. We also present a brief review of an emerging class of thermoelectric materials, electrically conducting polymers.

3.2.1 Carrier concentration optimization

In this section, we explain how the electron transport properties (S, σ, and κ_e) depend on carrier concentration, and emphasize the importance of optimizing the carrier concentration for the high power factor.

In Fig. 3.1(a), we show the dependence of the electron transport properties on the carrier concentration. For simplicity, we assume a constant value for the relaxation time and a simple parabolic band for the group velocity and the density of states in Eqs. (3.8)–(3.10).

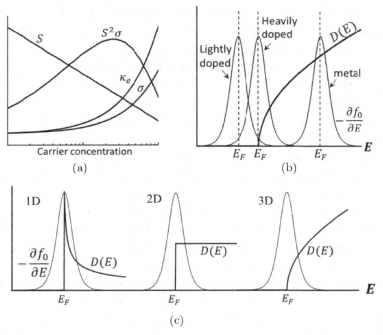

Figure 3.1. Dependence of electrical properties on (a and b) carrier concentration and (c) material dimensionality.

Figure 3.1(a) clearly shows that the Seebeck coefficient, electrical conductivity, and electronic thermal conductivity are interrelated via carrier concentration. As carrier concentration increases, the electrical conductivity and electronic thermal conductivity increase but the Seebeck coefficient decreases, implying that there is an optimal value of carrier concentration for the highest power factor. The dependence of the electrical conductivity and electronic thermal conductivity on carrier concentration is straightforward to understand; more the carriers exist, more the charges or heat transport. To help understand why the Seebeck coefficient changes with carrier concentration, we point out that the Seebeck coefficient is due to the asymmetry of electron transport parameters, such as the density of states about the Fermi level. The physical interpretation of the Seebeck coefficient is the average entropy $(E - E_F)/T$, that electrons carry (Eq. (3.9)). The electrons above the Fermi level carry the positive entropy, while those below the Fermi level carry the negative entropy, and the Seebeck coefficient is the average of the two quantities. Therefore, the nonzero Seebeck coefficient implies imbalance between contributions from the electrons above the Fermi level and below the Fermi level, which strongly depends on the Fermi level position. In Fig. 3.1(b), we plot the density of states of a simple parabolic band and the Fermi windows $(-\partial f_0/\partial E)$ for lightly doped, heavily doped, and metallic cases. In the metallic case, the Fermi level is deep inside the band where the density of states slowly increases with energy. Here, the contribution from the electrons above the Fermi level is largely compensated by the contribution from the electrons below the Fermi level, leading to a small Seebeck coefficient. In the lightly doped case, electrons below the Fermi level are prohibited by the band gap, leading to a large Seebeck coefficient. However, the carrier concentration is too low in this case, resulting in a low power factor. Therefore, carrier concentration is the primary factor determining the electron transport properties, and finding optimal carrier concentration is important to maximize the power factor.

In the aforementioned discussion, we assumed a parabolic band structure for simplicity, but many good thermoelectric materials have more complicated band structures near the band edge. For example, lead chalcogenides are thought to have a second heavy band[9] or connection between carrier pockets[10] at a certain energy level near the valence band edge. Both the second heavy hole band and the connection between carrier pockets make an abrupt change in the density of states. The abrupt change in the density of states implies that the band structure cannot be simply described by a single parameter, such as an effective mass. Hence, to find the optimal value of carrier concentration for such a complicated band structure, a wide range of carrier concentration should be carefully studied rather than projecting the transport properties from the measured data at a certain carrier concentration.

3.2.2 Band engineering

The idea of engineering an electronic band structure was initiated by Hicks and Dresselhaus.[5,6] They suggested that the power factor can be improved when a conventional thermoelectric material is fabricated in the form of 2D quantum-well superlattice or 1D nanowire. The quantum confinement creates sharp features in the density of states as shown in Fig. 3.1(c) that are more favorable for the high power factor as compared to that in the 3D structure, since they exhibit larger asymmetry with respect to the Fermi level. Mahan and Sofo mathematically showed that a delta function is the ideal form of the density of states and energy-dependent relaxation time to maximize the power factor,[11] although some broadening is needed for actual transport.[12,13] The delta function form of the density of states also leads to zero electronic thermal conductivity, and the energy exchange by electron transport can be a reversible process.[14]

The Hicks and Dresselhaus work stimulated experimental studies in the following decades. For the proof of concept, the 2D quantum-well structure ($PbTe/Pb_{1-x}Eu_xTe$) was studied and shown to have a larger value for the product of the Seebeck coefficient and carrier concentration (S^2n) than bulk PbTe by a factor of more than 2.[15,16] Further proof-of-concept studies were carried out with Bi nanowires,[17,18] which exhibit a semimetal to semiconductor transition due to the quantum confinement of electrons and holes. Another

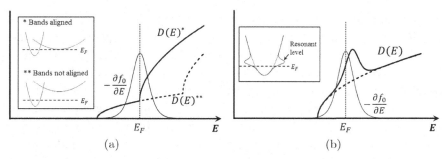

Figure 3.2. Effects of (a) the band alignment and (b) the resonant level on density-of-states. Schematic band structures are presented in the subfigures.

interesting approach relying on the low dimensionality is the alignment of bands.[19,20] The band engineering through superlattices[19] or alloys in nanowires[20] can make the second heavy band align with the first band as shown in Fig. 3.2(a). The alignment largely increases the density of states near the band edge by the contribution from the second band, leading to larger asymmetry in the density of states with respect to the band edge.

Subsequent efforts have shifted to search for sharp features and large density of states in 3D materials. One successful approach is adding resonant impurities to the conventional thermoelectric materials.[21] The resonant impurities form the impurity states, called resonant levels, inside bands. These impurity states are extended states rather than the localized states, which are usually seen in the case of normal impurities, enabling the electrons in the impurity states to contribute to charge transport.[22] The resonant level makes a hump in the density of states as schematically shown in Fig. 3.2(b), increasing the asymmetry of the density of states about the Fermi level and thereby enhancing the thermoelectric power factor. This explains large increases in the power factor of Tl-doped PbTe[21] and In-doped SnTe.[23]

The band alignment idea in low-dimensional materials shown in Fig. 3.2(a) found its analog in bulk $PbTe_{1-x}Se_x$.[9] The energy difference between the band edges of the first and the second bands in PbTe depends on temperature and the alloying ratio with PbSe. By alloying PbTe with a proper amount of PbSe, the temperature at which the first and second valence bands are aligned could be tuned to a practically important range near 500 K, leading to large ZT in this temperature range. The band alignment idea was also successfully applied to another material system, $Mg_2Si_{1-x}Sn_x$.[24]

3.2.3 Carrier-scattering control

In Eqs. (3.8) and (3.9), one remaining parameter we can control is the relaxation time, τ. There have been two approaches to control the relaxation time: one is making the relaxation time strongly depend on energy to increase the Seebeck coefficient and the other is increasing the relaxation time to increase carrier mobility and thereby electrical conductivity. The former approach is similar to the idea of the sharp density of states. Considering that the integrands in Eqs. (3.8)–(3.10) consist of the product of both relaxation time and density of states, the strongly energy-dependent relaxation time increases the asymmetry of electron transport about the Fermi level, in the same way as the sharp feature of the density of states does. The strongly energy-dependent relaxation time can be achieved by scattering low-energy electrons preferentially, known as electron filtering. Then, the Seebeck coefficient can be enhanced significantly owing to the largely asymmetric relaxation time about the Fermi level. The electron filtering was shown to improve the Seebeck coefficient in 2D superlattices[25,26] and nanograined PbTe[27] in which interfaces form energy barriers that preferentially scatter low-energy electrons.

Attempts have also been there to increase electrical conductivity by increasing the relaxation time of electrons. The idea of modulation doping, which has been widely used in microelectronics, was applied to bulk nanocomposite thermoelectric materials to enhance electrical conductivity.[28] In the modulation doping, carriers and dopants are spatially separated so that carriers experience fewer scattering events by dopants. The separation was achieved by encapsulating dopants in nanoparticles. The carriers from the dopants in nanoparticles are spilled to the host material due to the band bending near the interface between the host material and the nanoparticles. Since all dopants are encapsulated within the nanoparticles, the transport of carriers in the host material is less affected by the impurity scattering, resulting in high mobility and electrical conductivity. Antiresonant scattering by invisible dopants was also theoretically suggested.[29,30] The invisible dopants can be composed of a core–shell structure with appropriately designed potentials. The electron-scattering cross-section of the properly designed core–shell structures can be sharply reduced within a narrow energy window near the Fermi level. The small scattering cross-section sharply changing with energy contributes to the enhancement of the power factor in two different ways. The small scattering cross-section improves electron mobility and its large energy dependence increases the Seebeck coefficient as in the electron-filtering effect.

3.2.4 Polymer-based thermoelectrics

Conducting polymers have drawn much attention as potential thermoelectric materials in recent decades because of their several advantages over inorganic thermoelectric materials. Polymers are easy to process, cost-effective compared to inorganic thermoelectric materials, and do not require any rare or toxic elements such as Te and Pb which are commonly used in inorganic thermoelectric materials. Moreover, the thermal conductivity of polymers is usually around $0.4\,\mathrm{Wm^{-1}K^{-1}}$, even lower than that of the state-of-the-art inorganic thermoelectric materials. Despite these advantages, polymers exhibit a small power factor, limiting their ZT to be at least two orders of magnitude smaller than that of inorganic thermoelectric materials. Therefore, recent efforts for thermoelectric polymers have focused on improving their power factors. For example, optimizing carrier concentration is a key to achieving a high power factor in thermoelectric polymers, just as in inorganic semiconductors. In fact, carrier concentration optimization improved the power factor of poly(3,4-ethylenedioxythiophene) (PEDOT) by more than ten times.[31] Besides carrier concentration optimization, most power factor improvements have been achieved by two main strategies: (i) understanding and controlling the morphology of microstructures in polymers and (ii) developing polymer and inorganic nanomaterial

composites. Here we summarize recent studies in these two cate-
gories. A more comprehensive review for each kind of polymers can
be found in the literature.[32]

Unlike inorganic materials, polymers are not homogenous and,
thus, have certain morphology in their microstructures. This
inhomogeneity hampers electron transport; therefore, understanding and controlling the morphology is important for achieving high electrical conductivity and, thus, a high power factor.
As an example, one of the most promising thermoelectric
polymers, poly(3,4-ethylenedioxythiophene):poly(styrene sulfonate)
(PEDOT:PSS), exhibits phase segregation between the electrically
conducting PEDOT phase and the insulating PSS phase, confirmed by various analysis techniques, such as photoelectron spectroscopy[33–35] and scanning tunneling microscopy.[36,37] These studies
revealed that PEDOT:PSS consist of PEDOT-rich cores and PSS-
rich shells. Further studies revealed pancake-like PEDOT regions
separated by PSS lamella along the in-plane direction, explaining
the high electrical conductivity anisotropy between the in-plane and
out-of-plane directions.[38,39] In parallel to the efforts to understand
morphology, there were many attempts to control morphology in
order to improve the power factor. The treatment of PEDOT:PSS
with some hydrophilic solvents, such as dimethylsulfoxide (DMSO)
and ethylene glycol (EG), was found to improve electron mobility by
several orders of magnitudes.[40–42] The proposed mechanisms of the
electron mobility improvements are twofold: (i) the treatment with
the hydrophilic solvents partially removes the hydrophilic PSS phase
while conserving the hydrophobic PEDOT phase and (ii) the morphology is modified such that there are more connections between
PEDOT phases. A similar approach using DMSO and EG finally led
to the highest ZT of 0.42 among the polymer-based thermoelectric
materials so far.[43]

The polymer–inorganic nanomaterial composite was suggested to
improve the low power factor of polymers by taking advantage of the
high power factor of inorganic semiconductors. In early studies of
composites of electrically insulating polymers and carbon nanotubes
(CNTs), it was shown that the CNTs form an electrical network due

to the percolation effect and correspondingly the electrical conductivity of the composite is much improved from that of the intrinsic polymers.[44-46] Thermal conductivity, however, intriguingly remains low in the nanocomposite.[46] It was inferred from these experiments that the network between CNTs in the nanocomposites is electrically conducting but thermally insulating, ideal for thermoelectric applications. In a subsequent study, PEDOT:PSS instead of the electrically insulating polymers was used in a nanocomposite form with CNTs. The nanocomposite exhibits ZT of 0.02, an improvement by at least one order of magnitude over PEDOT:PSS that is not optimized for the carrier concentration.[47] In addition to the network formed between CNTs, the inclusion of CNTs can affect the morphology of polymers. A study of a polyaniline and CNT composite suggested that the presence of CNTs can cause the polymer chains to be well ordered by $\pi-\pi$ interactions between the CNTs and polymer chains during *in situ* polymerization reaction, resulting in improved electron mobility in the polymer phase.[48]

Other inorganic materials have also been incorporated into composites to improve the power factor. The inclusion of Bi_2Te_3 nanoparticles into PEDOT:PSS[49] and polyaniline[50] was shown to improve the power factor. Another noteworthy result was reported in the nanocomposite of PEDOT:PSS and Te nanorods.[51] A fundamental limit of polymer–inorganic nanocomposite according to the conventional theory for composite thermoelectric materials, is that the ZT of a composite material cannot exceed the ZT of any constituent material under the assumption that the transport properties in each constituent are not affected by thepresence of other constituents.[52] However, the nanocomposite of PEDOT:PSS and Te nanorods interestingly showed ZT larger than that of either constituents.[51] It was suggested in a subsequent study that the interface between Te nanorod and PEDOT:PSS plays an important role in such anomalous improvement in ZT, though the detailed mechanism has not been fully understood.[53]

The approaches devised for improving inorganic thermoelectric materials, which we discussed in the previous section, were applied to polymer and inorganic material composites. The electron-filtering effect was found in poly(3-hexylthiophene) (P3HT) and Bi_2Te_3 nanowire composites.[54] While the carrier concentration of the composite is similar to that of P3HT, the Seebeck coefficient and the electrical conductivity of the composite were significantly increased and decreased, respectively, compared to intrinsic P3HT, which may indicate the electron-filtering effect at the interface between P3HT and Bi_2Te_3 nanowires. Modulation doping was recently applied to the planar heterostructure of PEDOT:PSS and undoped Si.[55] Both undoped Si and PEDOT:PSS exhibit a low power factor, but the reasons for the low power factor are different; undoped Si has high electron mobility with low carrier concentration, whereas PEDOT:PSS has low electron mobility with high carrier concentration. In the heterostructure of undoped Si and PEDOT:PSS, however, many charge carriers are spilled from the PEDOT:PSS to the undoped Si,and the mobility of these carriers maintains the high mobility of the undoped Si, leading to an increase in the power factor.

3.3 Phonon Transport

In addition to the improvements made to the power factor, many of the significant enhancements of ZT in recent years were achieved by reducing phonon thermal conductivity.[7,56] In this section, we review past works about reducing thermal conductivity, and the several different points of view explaining this reduction.

In thermoelectric materials, there are three heat conduction mechanisms degrading ZT and device efficiency, including heat conduction by phonons and electrons, and the bipolar effect. Bipolar heat conduction is due to the motion of electrons and holes together that carry energy across the bandgap directly from the high to the low temperature regions. It is significant at high temperatures when the number of thermally excited minority charge carriers becomes comparable to that of the majority charge carriers. Normal electronic contributions to thermal conductivity can be related to the electrical conductivity through the Wiedemann–Franz law, with the Lorenz number weakly dependent on the carrier concentration. The major heat carriers in thermoelectric materials are usually phonons and

here we focus our discussion on phonon transport. From the Boltzmann equation with the relaxation time approximation, the phonon
thermal conductivity, κ, is given as

$$\kappa = \frac{1}{3} \sum_{\lambda\mathbf{k}} C_{\lambda\mathbf{k}} v_{\lambda\mathbf{k}} \Lambda_{\lambda\mathbf{k}}, \tag{3.13}$$

where $C_{\lambda\mathbf{k}}$, $v_{\lambda\mathbf{k}}$, and $\Lambda_{\lambda\mathbf{k}}$ are specific heat, group velocity, and mean
free path, respectively, of a phonon mode with the polarization λ
and wavevector $\mathbf{k}$. To reduce the thermal conductivity, we need to
control the three parameters as aforementioned. For most of the time
the specific heat is fixed to the Dulong–Petit limit since most thermoelectric materials have a small Debye temperature for low intrinsic
thermal conductivity. On the other hand, the remaining two parameters, the group velocity and the phonon mean free path, can be
controlled by engineering a phonon band structure and employing
extrinsic scattering mechanisms, respectively. In most of the cases
where thermal conductivity was effectively reduced, the reduction of
the phonon mean free path has been considered a main reason. However, the change in group velocity has also been regarded as another
important contribution in some cases, and a clear understanding of
the thermal conductivity reduction is still lacking. Therefore, the
past studies on phonon transport, unlike those on electron transport,
will be categorized depending on material forms rather than specific
mechanisms of the thermal conductivity reduction.

Before discussing on how thermal conductivity can be reduced,
we would like to emphasize that the broad spectrum of phonons is
involved in thermal transport. Equation (3.13) is often simplified to
$\kappa = (1/3)Cv\Lambda$ by assuming an averaged phonon mode. However, this
simplified expression loses the critical feature of phonon transport
that the mean free path spans over several orders of magnitudes.
Unlike electron transport where only small carrier pockets in the
Brillouin zone are important, in the case of phonon transport, the
broad spectrum of phonons is excited and carries thermal energy.
The broad spectrum of phonons leads to a large range of mean free
path because the strength of phonon-scattering mechanisms strongly
depends on the phonon frequency (ω)). For example, the most

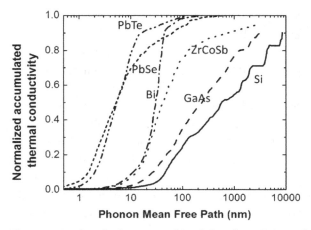

Figure 3.3. First principle calculation results of the phonon mean free path distributions in single crystalline thermoelectric materials, Si,[60] GaAs,[61] ZrCoSb,[62] PbTe,[63] PbSe,[64] and Bi.[65]

dominant phonon-scattering mechanism in usual cases, three-phonon scattering, exhibits a scattering rate approximately proportional to the second power of phonon frequency.[57] Therefore, the phonon mean free path varies largely with its frequency, resulting in a wide range of mean free path. The wide span of mean free path was recently confirmed by numerical and experimental works.[58–66] In Fig. 3.3, we have shown the broad distribution of mean free path in many thermoelectric materials, calculated from first principles.[60–63,65] This broadly distributed mean free path is the basis of the recent thermal conductivity reduction we will review in Sections 3.3 and 3.4.

3.3.1 Increasing phonon-scattering rate by disorder in alloys

Since it was first suggested in the 1950s,[67] alloying has been widely used to reduce thermal conductivity. The idea is to introduce the disorder of atomic masses and force constants into the crystalline phase to scatter phonons. By adding isoelectric elements in the same column of the periodic table, such as adding Ge to Si, phonons are effectively scattered while the electrical properties are not significantly affected. Alloying has been effective for reducing thermal conductivity, and the most of the good thermoelectric materials have been developed in an alloy form, such as $Bi_2Te_{3-x}Se_x$, and $PbTe_{1-x}Se_x$.

Even though alloying has been successful for long time, the phonon transport in alloys is still not well understood. The substitution atom in an alloy is often described as a point defect, and the phonon scattering by the point defect can be understood with the Rayleigh scattering model which exhibits the scattering rate proportional to the fourth power of the phonon frequency. The Rayleigh scattering model was successful in reproducing the measured thermal conductivity values for $Si_{1-x}Ge_x$ alloy,[68,69] but there are still several questions which need to be answered. For example, the Rayleigh scattering model is valid only when phonon wavelength is long compared to the characteristic size of the disorder. Therefore, it may not be valid for short wavelength phonons. Also, when the alloying ratio is large, for example of $Si_{50}Ge_{50}$, the defect should be described as a cluster rather than a point. Moreover, with such a high concentration of defects, a correlation among many defect scatterings possibly exists, leading to phonon localization.[70]

3.3.2 Increasing phonon-scattering rate and engineering phonon band by rattling atoms in cages

Some rare earth atoms are known to exhibit large thermal vibrations (rattle) when they are in the cage of host material that are large enough to accommodate the rare earth atom.[71] Since such a large thermal vibration of an atom in the cage can induce strong scattering of phonons in the host material, Slack suggested filling the cages of skutterudite with the rare earth atoms to reduce thermal conductivity.[72] This approach has been shown to effectively reduce thermal conductivity in skutterudite[73-75] and clathrate[76] material systems, but the microscopic mechanism for this reduction is still elusive and under debates. The original idea of the rattler approach was that the rattler vibrates *incoherently* with host atoms and scatters the phonons in host materials.[77,78] However, high-resolution neutron-scattering measurements show that the rattler vibrates *coherently* with host atoms, generating phonon modes different from the original modes of host materials.[79–81] The phonon modes modified by the rattlers contain many avoided crossings between phonon branches and the group velocity is correspondingly reduced. Both incoherent and coherent vibration pictures can explain the reduction in thermal conductivity, but the underlying mechanism is different; the former one increases the phonon-scattering rate, whereas the latter one decreases group velocity by modifying the phonon band structure. Besides these mechanisms, the alloy effect can be another possible explanation for the reduced thermal conductivity of partially filled cage cases. When cages are partially filled, the disorder of filled and unfilled cages can scatter phonons in many alloys.[82,83]

3.3.3 Increasing phonon-scattering rate and engineering phonon band in low-dimensional materials

The nanostructure approach to create improved thermoelectric materials suggested by Hicks and Dresselhaus[5,6] implied that in addition to the electron properties, the phonon properties can be also engineered. They assumed that phonon mean free path is limited to the characteristic size of nanostructure due to strong boundary scattering. Therefore, thermal conductivity is simply reduced to $(1/3)CvL$, called the Casimir limit, where the L is the characteristic size of nanostructure. This leads to the investigation of thermal conductivity in low-dimensional materials. The thermal conductivity lower than that of the bulk material was observed in superlattices,[84–86] Si nanowires[87] and thin Si membranes with high density of holes.[88] In particular, the thermal conductivity of the Si nanowire[87] and thin Si membranes with high density of holes[88] decrease with decreasing the characteristic size of nanostructure, confirming that the characteristic size of nanostructure is a decisive parameter for thermal conductivity in the Casimir limit.

In addition to the size of nanostructure, surface roughness was found another decisive parameter of reducing thermal conductivity. Extremely low thermal conductivity, even below the Casimir limit, was reported in the Si nanowire having a roughened surface.[89,90] The Si nanowire with the roughened surface[89,90] exhibits much lower thermal conductivity than the smooth Si nanowire[87] with the same diameter. The origin of the extremely low thermal conductivity even below the Casimir limit is not still clear. Several numerical simulations draw different conclusions about whether the low thermal conductivity is solely attributed to roughness or not.[91,92] Even though the microscopic mechanism about the low thermal conductivity of the rough Si nanowires is not fully explained, surface roughness is clearly a key factor for reducing thermal conductivity, according to the recent quantitative analysis of experimental data.[93]

The above discussion is mostly about the phonon mean free path, but the group velocity in Eq. (3.13) can also be altered by introducing nanostructures. For example, if a nanowire has a smaller diameter than the intrinsic phonon mean free path and its surface is smooth enough to cause specular phonon scatterings, phonons can reflect from the boundaries of the nanowire without losing phase information. This coherent phonon transport can alter the phonon band structure and change the group velocity. A similar effect is expected in a superlattice with a smooth interface and short layer spacing. In this case, the periodic change in acoustic properties causes Bragg reflection which results in bending of the dispersion relation and opening of the phonon band gaps. The conditions for coherent phonon transport, which are smooth interfaces and the small size of nanostructure, seem to be stringent. However, recalling that the broad spectrum of phonons carries heat, the conditions are not unattainable for low-frequency phonons. The low-frequency phonons have a long intrinsic mean free path and experience specular scattering rather than diffuse scattering at interfaces.[94] The recent measurements of thermal conductivity suggest that coherent phonon transport possibly exists in the Ge–Si core–shell nanowires[95] and phononic structure.[96,97] More direct evidence of the coherent phonon transport was reported using the superlattice structure with the various number of periods.[98]

3.3.4 Increasing phonon-scattering rate in nanocomposite

After the successful demonstration of the low thermal conductivity in the low-dimensional materials, nanostructures were employed to bulk materials to reduce thermal conductivity. Two main strategies were followed, namely embedding nanoparticles and making many grain boundaries. Both strategies were pursued in an effort to reduce the thermal conductivity below than that found in alloys. Before the nanostructure approach was applied, the thermal conductivity in alloys was considered the lowest one achievable while keeping good electrical properties. However, as discussed in the previous sections, the rates of intrinsic three-phonon scattering and point defect scattering are proportional to the second and fourth powers of phonon frequency, respectively, implying that low-frequency phonons are not effectively scattered. Several bulk nanostructured materials were developed to more effectively scatter low-frequency phonons as discussed in the following.

The idea of embedding nanoparticles was demonstrated using the $In_{0.53}Ga_{0.47}As$ alloy with ErAs nanoparticles embedded.[99] The ErAs nanoparticles with the size of few nanometers were proven to effectively scatter low-frequency phonons, and thermal conductivity could be reduced below the alloy limit. In addition, the scattering by grain boundaries, similar to the strong boundary scattering shown in the low-dimensional materials, was also exploited. The nanocomposite $Bi_2Sb_{1.4}Te_{0.6}$ and $Si_{80}Ge_{20}$ alloys, consisting of many nanograins, were developed by easily scalable ball-milling and hot-pressing processes, and exhibit the improvement of ZT by 40–50% due to the reduction of thermal conductivity.[56,100] In the nanocomposite alloys, phonon scattering is augmented for the entire range of phonon spectrum: high-frequency phonons scattered by the alloy disorder and low-frequency phonons scattered by the rough grain boundaries and many defects formed during the ball-milling process.[101] The idea of scattering the whole spectrum of phonons was also applied to heavily Na-doped PbTe with nano-precipitates embedded endotaxially to minimally affect electron mobility, resulting in the highest $ZT \sim 2.2$ among many bulk nanostructured materials.[7]

Another interesting bulk nanostructured material is $AgSbTe_2$. It was discovered that $AgSbTe_2$ have exceptionally low thermal conductivity in 1950s,[102] but its low thermal conductivity is not clearly understood. The strong anharmonicity was first considered a main reason for the low thermal conductivity.[103] However, the natural nanostructure in $AgSbTe_2$, which are due to many competing phases with the similar energy levels, was recently suggested to be the dominant factor.[104] More studies are needed to confirm that the natural nanostructure actually causes the low thermal conductivity of $AgSbTe_2$, but the natural nanostructures might be promising as they are thermodynamically stable unlike many other artificial nanostructures.

3.4 Summary

So far, we have reviewed how applying nanostructures has led to the recent significant enhancement of ZT. Electron band engineering through the low-dimensional materials has successfully demonstrated that nanostructures can improve the power factor. Subsequently, the conventional bulk thermoelectric materials were modified so that they have the similar features to low-dimensional materials such as the sharp density of states. The increase of ZT through these modifications was mostly limited to lead chalcogenides, and extending the current approaches to other material systems is a future challenge. Another promising future strategy to improve the power factor further is to control carrier scattering. In contrast to the band engineering, power factor improvement by the control of carrier scattering has been demonstrated in only few cases at present.[28] The remarkable improvements in ZT of polymer-based thermoelectric materials in recent years also indicate another direction. Insights from the past works in inorganic thermoelectric materials may lead to further improvements of polymer-based thermoelectric materials.

Reducing thermal conductivity by utilizing nanostructures has been the most successful approach used to improve ZT in recent years. The nanostructures combined with the alloying effect can effectively scatter the entire spectrum of phonons. As a result, the thermal conductivity of the nanostructured thermoelectric materials at present is somewhat close to the amorphous limit, which had long been thought as a theoretical lower limit of thermal conductivity.[105,106] However, it seems that there is still room for reducing thermal conductivity further. The recent report of the ultralow thermal conductivity in WSe_2 shows that the thermal conductivity below the amorphous limit is possible.[107] Furthermore, a clear understanding about phonon transport in the thermoelectric materials is still lacking. The debates about the coherent and incoherent phonon transport in the low-dimensional materials are noteworthy in a sense that the current approach to reduce thermal conductivity relies only on the incoherent picture, such as increased scattering by defects and grain boundaries. Phonon band engineering by the coherent phonon transport might be possibly a new opportunity to further reduce the thermal conductivity.

Acknowledgments

S.L. would like to thank J. Cuffe, B. Liao, and D. Lee for their comments on the manuscript. This work was partially supported by "Solid State Solar-Thermal Energy Conversion Center (S3TEC)," an Energy Frontier Research Center funded by the U.S. Department of Energy, Office of Science, Office of Basic Energy Sciences under the award number DE-SC0001299/DE-FG02-09ER46577 (G.C.), and partially supported by DoD AFOSR MURI via Ohio State Univ. (S.L.). S.L. also acknowledges support from a Samsung scholarship.

References

1. L. E. Bell, Cooling, heating, generating power, and recovering waste heat with thermoelectric systems. *Science*, **321**, 1457 (2008).
2. D. Kraemer, B. Poudel, H.-P. Feng, J. C. Caylor, B. Yu, X. Yan, Y. Ma, X. Wang, D. Wang, A. Muto, K. McEnaney, M. Chiesa, Z. Ren,

and G. Chen, High-performance flat-panel solar thermoelectric generators with high thermal concentration. *Nat. Mater.*, **10**, 532 (2011).

3. C. B. Vining, An inconvenient truth about thermoelectrics. *Nat. Mater.*, **8**, 83 (2009).

4. G. S. Nolas, J. Sharp, and H. J. Goldsmid, *Thermoelectrics: Basic Principles and New Materials Developments,*Springer: New York (2001).

5. L. D. Hicks and M. S. Dresselhaus, Effect of quantum-well structures on the thermoelectric figure of merit. *Phys. Rev. B*, **47**, 12727 (1993).

6. L. D. Hicks and M. S. Dresselhaus, Thermoelectric figure of merit of a one-dimensional conductor. *Phys. Rev. B*, **47**, 16631 (1993).

7. K. Biswas, J. He, I. D. Blum, C.-I. Wu, T. P. Hogan, D. N. Seidman, V. P. Dravid, and M. G. Kanatzidis, High-performance bulk thermoelectrics with all-scale hierarchical architectures. *Nature*, **489**, 414 (2012).

8. G. Chen, *Nanoscale Energy Transport and Conversion: A Parallel Treatment of Electrons, Molecules, Phonons, and Photons*, Oxford University Press: USA, 2005.

9. Y. Pei, X. Shi, A. LaLonde, H. Wang, L. Chen, and G. J. Snyder, Convergence of electronic bands for high performance bulk thermoelectrics. *Nature*, **473**, 66 (2011).

10. D. Parker, X. Chen, and D. J. Singh, High three-dimensional thermoelectric performance from low-dimensional bands. *Phys. Rev. Lett.*, **110**, 146601 (2013).

11. G. Mahan and J. Sofo, The best thermoelectric. *Proc. Natl Acad. Sci.*, *Proc. Natl Acad. Sci.*, **7436** (1996).

12. J. Zhou, R. Yang, G. Chen, and M. S. Dresselhaus, Optimal bandwidth for high efficiency thermoelectrics. *Phys. Rev. Lett.*, **107**, 226601 (2011).

13. C. Jeong, R. Kim, and M. S. Lundstrom, On the best bandstructure for thermoelectric performance: A Landauer perspective. *J. Appl. Phys.*, **111** (2012).

14. T. E. Humphrey and H. Linke, Reversible Thermoelectric Nanomaterials. *Phys. Rev. Lett.*, **94**, 096601 (2005).

15. T. Koga, T. C. Harman, S. B. Cronin, and M. S. Dresselhaus, Mechanism of the enhanced thermoelectric power in (111)-oriented n-type $PbTe/Pb_{1-x}Eu_xTe$ multiple quantum wells. *Phys. Rev. B*, **60**, 14286 (1999).

16. L. D. Hicks, T. C. Harman, X. Sun, and M. S. Dresselhaus, Experimental study of the effect of quantum-well structures on the thermoelectric figure of merit. *Phys. Rev. B*, **53**, R10493 (1996).

17. Y.-M. Lin, X. Sun, and M. S. Dresselhaus, Theoretical investigation of thermoelectric transport properties of cylindrical Bi nanowires. *Phys. Rev. B*, **62**, 4610 (2000).
18. J. P. Heremans, C. M. Thrush, D. T. Morelli, and M.-C. Wu, Thermoelectric power of bismuth nanocomposites. *Phys. Rev. Lett.*, **88**, 216801 (2002).
19. T. Koga, X. Sun, S. B. Cronin, and M. S. Dresselhaus, Carrier pocket engineering to design superior thermoelectric materials using GaAs/AlAs superlattices. *Appl. Phys. Lett.*, **73**, 2950 (1998).
20. O. Rabina, Y.-M. Lin, and M. S. Dresselhaus, Anomalously high thermoelectric figure of merit in $Bi_{1-x}Sb_x$ nanowires by carrier pocket alignment. *Appl. Phys. Lett.*, **79**, 81 (2001).
21. J. Heremans, V. Jovovic, E. Toberer, A. Saramat, K. Kurosaki, A. Charoenphakdee, S. Yamanaka, and G. Snyder, Enhancement of thermoelectric efficiency in PbTe by distortion of the electronic density of states. *Science*, **321**, 554 (2008).
22. J. P. Heremans, B. Wiendlocha, and A. M. Chamoire, Resonant levels in bulk thermoelectric semiconductors. *Ener. Environ. Sci.*, **5**, 5510 (2012).
23. Q. Zhang, B. Liao, Y. Lan, K. Lukas, W. Liu, K. Esfarjani, C. Opeil, D. Broido, G. Chen, and Z. Ren, High thermoelectric performance by resonant dopant indium in nanostructured SnTe. *Proc. Natl Acad. Sci.*, **110**, 13261 (2013).
24. W. Liu, X. Tan, K. Yin, H. Liu, X. Tang, J. Shi, Q. Zhang, and C. Uher, Convergence of conduction bands as a means of enhancing thermoelectric performance of n-type $Mg_2Si_{1-x}Sn_x$ solid solutions. *Phys. Rev. Lett.*, **108**, 166601 (2012).
25. D. Vashaee and A. Shakouri, Improved thermoelectric power factor in metal-based superlattices. *Phys. Rev. Lett.*, **92**, 106103 (2004).
26. J. M. O. Zide, D. Vashaee, Z. X. Bian, G. Zeng, J. E. Bowers, A. Shakouri, and A. C. Gossard, Demonstration of electron filtering to increase the Seebeck coefficient in $In_{0.53}Ga_{0.47}As/In_{0.53}Ga_{0.28}Al_{0.19}As$ superlattices. *Phys. Rev. B*, **74**, 205335 (2006).
27. J. Heremans, C. Thrush, and D. Morelli, Thermopower enhancement in lead telluride nanostructures. *Phys. Rev. B*, **70**, 115334 (2004).
28. M. Zebarjadi, G. Joshi, G. Zhu, B. Yu, A. Minnich, Y. Lan, X. Wang, M. Dresselhaus, Z. Ren, and G. Chen, Power factor enhancement by modulation doping in bulk nanocomposites. *Nano Lett.*, **11**, 2225 (2011).
29. B. Liao, M. Zebarjadi, K. Esfarjani, and G. Chen, Cloaking core–shell nanoparticles from conducting electrons in solids. *Phys. Rev. Lett.*, **109**, 126806 (2012).

30. M. Zebarjadi, B. Liao, K. Esfarjani, M. Dresselhaus, and G. Chen, Enhancing the thermoelectric power factor by using invisible dopants. *Adv. Mater.*, **25**, 1577 (2013).
31. O. Bubnova, Z. U. Khan, A. Malti, S. Braun, M. Fahlman, M. Berggren, and X. Crispin, Optimization of the thermoelectric figure of merit in the conducting polymer poly(3,4-ethylenedioxythiophene). *Nat. Mater.*, **10**, 429 (2011).
32. Y. Du, S. Z. Shen, K. Cai, and P. S. Casey, Research progress on polymer–inorganic thermoelectric nanocomposite materials. *Prog. Polym. Sci.*, **37**, 820 (2012).
33. X. Crispin, S. Marciniak, W. Osikowicz, G. Zotti, A. W. D. van der Gon, F. Louwet, M. Fahlman, L. Groenendaal, F. De Schryver, and W. R. Salaneck, Conductivity, morphology, interfacial chemistry, and stability of poly(3,4-ethylene dioxythiophene)–poly(styrene sulfonate): A photoelectron spectroscopy study. *J. Polym. Sci. B: Polym. Phys.*, **41**, 2561 (2003).
34. G. Greczynski, T. Kugler, M. Keil, W. Osikowicz, M. Fahlman, and W. R. Salaneck, Photoelectron spectroscopy of thin films of PEDOT–PSS conjugated polymer blend: A mini-review and some new results. *J. Electron Spectrosc. Relat. Phenomena*, **121**, 1 (2001).
35. G. Greczynski, T. Kugler, and W. R. Salaneck, Characterization of the PEDOT-PSS system by means of X-ray and ultraviolet photoelectron spectroscopy. *Thin Solid Films*, **354**, 129 (1999).
36. M. Kemerink, S. Timpanaro, M. M. de Kok, E. A. Meulenkamp, and F. J. Touwslager, Three-Dimensional Inhomogeneities in PEDOT:PSS Films. *J. Phys. Chem. B*, **108**, 18820 (2004).
37. S. Timpanaro, M. Kemerink, F. J. Touwslager, M. M. De Kok, and S. Schrader, Morphology and conductivity of PEDOT/PSS films studied by scanning-tunneling microscopy. *Chem. Phys. Lett.*, **394**, 339 (2004).
38. A. M. Nardes, M. Kemerink, R. A. J. Janssen, J. A. M. Bastiaansen, N. M. M. Kiggen, B. M. W. Langeveld, A. J. J. M. van Breemen, and M. M. de Kok, Microscopic understanding of the anisotropic conductivity of PEDOT:PSS thin films. *Adv. Mater.*, **19**, 1196 (2007).
39. U. Lang, E. Müller, N. Naujoks, and J. Dual, Microscopical investigations of PEDOT:PSS thin films. *Adv. Funct. Mater.*, **19**, 1215 (2009).
40. J. Ouyang, Q. Xu, C.-W. Chu, Y. Yang, G. Li, and J. Shinar, On the mechanism of conductivity enhancement in poly(3,4-ethylenedioxythiophene):poly(styrene sulfonate) film through solvent treatment. *Polymer*, **45**, 8443 (2004).

41. C. Liu, B. Lu, J. Yan, J. Xu, R. Yue, Z. Zhu, S. Zhou, X. Hu, Z. Zhang, and P. Chen, Highly conducting free-standing poly(3,4-ethylenedioxythiophene)/poly(styrenesulfonate) films with improved thermoelectric performances. *Synthetic Metals*, **160**, 2481 (2010).

42. D. Alemu, H.-Y. Wei, K.-C. Ho, and C.-W. Chu, Highly conductive PEDOT:PSS electrode by simple film treatment with methanol for ITO-free polymer solar cells. *Ener. Environ. Sci.*, **5**, 9662 (2012).

43. G. H. Kim, L. Shao, K. Zhang, and K. P. Pipe, Engineered doping of organic semiconductors for enhanced thermoelectric efficiency. *Nat. Mater.*, **12**, 719 (2013).

44. E. Kymakis and G. A. J. Amaratunga, Electrical properties of single-wall carbon nanotube-polymer composite films. *J. Appl. Phys.*, **99** (2006).

45. O. Meincke, D. Kaempfer, H. Weickmann, C. Friedrich, M. Vathauer, and H. Warth, Mechanical properties and electrical conductivity of carbon-nanotube filled polyamide-6 and its blends with acrylonitrile/butadiene/styrene. *Polymer*, **45**, 739 (2004).

46. C. Yu, Y. S. Kim, D. Kim, and J. C. Grunlan, Thermoelectric behavior of segregated-network polymer nanocomposites. *Nano Lett.*, **8**, 4428 (2008).

47. D. Kim, Y. Kim, K. Choi, J. C. Grunlan, and C. Yu, Improved thermoelectric behavior of nanotube-filled polymer composites with poly(3,4-ethylenedioxythiophene) Poly(styrenesulfonate). *ACS Nano*, **4**, 513 (2009).

48. Q. Yao, L. Chen, W. Zhang, S. Liufu, and X. Chen, Enhanced thermoelectric performance of single-walled carbon nanotubes/polyaniline hybrid nanocomposites. *ACS Nano*, **4**, 2445 (2010).

49. B. Zhang, J. Sun, H. E. Katz, F. Fang, and R. L. Opila, Promising thermoelectric properties of commercial PEDOT:PSS materials and their Bi_2Te_3 powder composites. *ACS Appl, Mater. Interfaces*, **2**, 3170 (2010).

50. N. Toshima, M. Imai, and S. Ichikawa, Organic–inorganic nanohybrids as novel thermoelectric materials: Hybrids of polyaniline and bismuth(III) telluride nanoparticles. *J. Electron. Mater.*, **40**, 898 (2011).

51. K. C. See, J. P. Feser, C. E. Chen, A. Majumdar, J. J. Urban, and R. A. Segalman, Water-processable polymer-nanocrystal hybrids for thermoelectrics. *Nano Lett.*, **10**, 4664 (2010).

52. D. J. Bergman and O. Levy, Thermoelectric properties of a composite medium. *J. Appl. Phys.*, **70**, 6821 (1991).

53. N. E. Coates, S. K. Yee, B. McCulloch, K. C. See, A. Majumdar, R. A. Segalman, and J. J. Urban, Effect of interfacial properties on

polymer–nanocrystal thermoelectric transport. *Adv. Mater.*, **25**, 1629 (2013).

54. M. He, J. Ge, Z. Lin, X. Feng, X. Wang, H. Lu, Y. Yang, and F. Qiu, Thermopower enhancement in conducting polymer nanocomposites via carrier energy scattering at the organic–inorganic semiconductor interface. *Ener. Environ. Sci.*, **5**, 8351 (2012).

55. D. Lee, C. A. Kuryak, C. E. Carlton, S. Lee, Y. Hu, G. Chen, and Y. Shao-Horn, *Material Research Society Fall Meeting*, Boston, MA, USA (2013).

56. B. Poudel, Q. Hao, Y. Ma, Y. Lan, A. Minnich, B. Yu, X. Yan, D. Wang, A. Muto, D. Vashaee, X. Chen, J. Liu, M. Dresselhaus, G. Chen, and Z. Ren, High-thermoelectric performance of nanostructured bismuth antimony telluride bulk alloys. *Science*, **320**, 634 (2008).

57. C. Herring, Role of low-energy phonons in thermal conduction. *Phys. Rev.*, **95**, 954 (1954).

58. A. J. Minnich, J. A. Johnson, A. J. Schmidt, K. Esfarjani, M. S. Dresselhaus, K. A. Nelson, and G. Chen, Thermal conductivity spectroscopy technique to measure phonon mean free paths. *Phys. Rev. Lett.*, **107**, 095901 (2011).

59. A. S. Henry and G. Chen, Spectral phonon transport properties of silicon based on molecular dynamics simulations and lattice dynamics. *J. Comput. Theor. Nanosci.*, **5**, 141 (2008).

60. K. Esfarjani, G. Chen, and H. T. Stokes, Heat transport in silicon from first-principles calculations. *Phys. Rev. B*, **84**, 085204 (2011).

61. T. Luo, J. Garg, J. Shiomi, K. Esfarjani, and G. Chen, Gallium arsenide thermal conductivity and optical phonon relaxation times from first-principles calculations. *Europhys. Lett.*, **101**, 16001 (2013).

62. J. Shiomi, K. Esfarjani, and G. Chen, Thermal conductivity of half-Heusler compounds from first-principles calculations. *Phys. Rev. B*, **84**, 104302 (2011).

63. T. Shiga, J. Shiomi, J. Ma, O. Delaire, T. Radzynski, A. Lusakowski, K. Esfarjani, and G. Chen, Microscopic mechanism of low thermal conductivity in lead telluride. *Phys. Rev. B*, **85**, 155203 (2012).

64. Z. Tian, J. Garg, K. Esfarjani, T. Shiga, J. Shiomi, and G. Chen, Phonon conduction in PbSe, PbTe, and $PbTe_{1-x}Se_x$ from first-principles calculations. *Phys. Rev. B*, **85**, 184303 (2012).

65. S. Lee, K. Esfarjani, J. Mendoza, M. S. Dresselhaus, and G. Chen, Lattice thermal conductivity of Bi, Sb, and Bi–Sb alloy from first principles. *Phys. Rev. B*, **89**, 085206 (2014).

66. S. Lee, K. Esfarjani, T. Luo, J. Zhou, Z. Tian, and G. Chen, Resonant bonding leads to low lattice thermal conductivity. *Nat. Commun.*, **5**, 3525 (2014).

67. A. F. Ioffe, *Semiconductor Thermoelements and Thermoelectric Cooling*, Infosearch: London, 1957.
68. B. Abeles, Lattice thermal conductivity of disordered semiconductor alloys at high temperatures. *Phys. Rev.*, **131**, 1906 (1963).
69. J. Garg, N. Bonini, B. Kozinsky, and N. Marzari, Role of disorder and anharmonicity in the thermal conductivity of silicon–germanium alloys: Afirst-principles study. *Phys. Rev. Lett.*, **106**, 045901 (2011).
70. I. Savić, N. Mingo, and D. A. Stewart, Phonon transport in isotope-disordered carbon and boron-nitride nanotubes: Is localization observable?. *Phys. Rev. Lett.*, **101**, 165502 (2008).
71. D. J. Braun and W. Jeitschko, Preparation and structural investigations of antimonides with the $LaFe_4P_{12}$ structure. *J. Less Common Metals*, **72**, 147 (1980).
72. G. A. Slack and V. G. Tsoukala, Some properties of semiconducting $IrSb_3$. *J. Appl. Phys.*, **76**, 1665 (1994).
73. B. C. Sales, D. Mandrus, and R. K. Williams, Filled skutterudite antimonides: A new class of thermoelectric materials. *Science*, **272**, 1325 (1996).
74. G. S. Nolas, M. Kaeser, R. T. Littleton, and T. M. Tritt, High figure of merit in partially filled ytterbium skutterudite materials. *Appl. Phys. Lett.*, **77**, 1855 (2000).
75. G. S. Nolas, J. L. Cohn, and G. A. Slack, Effect of partial void filling on the lattice thermal conductivity of skutterudites. *Phys. Rev. B*, **58**, 164 (1998).
76. G. S. Nolas, J. L. Cohn, G. A. Slack, and S. B. Schujman, Semiconducting Ge clathrates: Promising candidates for thermoelectric applications. *Appl. Phys. Lett.*, **73**, 178 (1998).
77. V. Keppens, D. Mandrus, B. C. Sales, B. C. Chakoumakos, P. Dai, R. Coldea, M. B. Maple, D. A. Gajewski, E. J. Freeman, and S. Bennington, Localized vibrational modes in metallic solids. *Nature*, **395**, 876 (1998).
78. B. C. Sales, D. Mandrus, B. C. Chakoumakos, V. Keppens, and J. R. Thompson, Filled skutterudite antimonides:Electron crystals and phonon glasses. *Phys. Rev. B*, **56**, 15081 (1997).
79. M. Christensen, A. B. Abrahamsen, N. B. Christensen, F. Juranyi, N. H. Andersen, K. Lefmann, J. Andreasson, C. R. H. Bahl, and B. B. Iversen, Avoided crossing of rattler modes in thermoelectric materials. *Nat. Mater.*, **7**, 811 (2008).
80. M. M. Koza, M. R. Johnson, R. Viennois, H. Mutka, L. Girard, and D. Ravot, Breakdown of phonon glass paradigm in La- and Ce-filled Fe_4Sb_{12} skutterudites. *Nat. Mater.*, **7**, 805 (2008).

81. M. Zebarjadi, K. Esfarjani, J. Yang, Z. F. Ren, and G. Chen, Effect of filler mass and binding on thermal conductivity of fully filled skutterudites. *Phys. Rev. B*, **82**, 195207 (2010).

82. G. P. Meisner, D. T. Morelli, S. Hu, J. Yang, and C. Uher, Structure and lattice thermal conductivity of fractionally filled skutterudites: Solid solutions of fully filled and unfilled end members. *Phys. Rev. Lett.*, **80**, 3551 (1998).

83. S. Q. Bai, X. Shi, and L. D. Chen, Lattice thermal transport in $Ba_xRE_yCo_4Sb_{12}$ (RE=Ce, Yb, and Eu) double-filled skutterudites. *Appl. Phys. Lett.*, **96**, 202102 (2010).

84. R. Venkatasubramanian, E. Siivola, T. Colpitts, and B. O'Quinn, Thin-film thermoelectric devices with high room-temperature figures of merit. *Nature*, **413**, 597 (2001).

85. T. Borca-Tasciuc, D. W. Song, J. R. Meyer, I. Vurgaftman, M.-J. Yang, B. Z. Nosho, L. J. Whitman, H. Lee, R. U. Martinelli, G. W. Turner, M. J. Manfra, and G. Chen, Thermal conductivity of $AlAs_{0.07}Sb_{0.93}$ and $Al_{0.9}Ga_{0.1}As_{0.07}Sb_{0.93}$ alloys and $(AlAs)_1/(AlSb)_{11}$ digital-alloy superlattices. *J. Appl. Phys.*, **92**, 4994 (2002).

86. B. Yang, J. L. Liu, K. L. Wang, and G. Chen, Simultaneous measurements of Seebeck coefficient and thermal conductivity across superlattice. *Appl. Phys. Lett.*, **80**, 1758 (2002).

87. D. Li, Y. Wu, P. Kim, L. Shi, P. Yang, and A. Majumdar, Thermal conductivity of individual silicon nanowires. *Appl. Phys. Lett.*, **83**, 2934 (2003).

88. J. Tang, H.-T. Wang, D. H. Lee, M. Fardy, Z. Huo, T. P. Russell, and P. Yang, Holey silicon as an efficient thermoelectric material. *Nano Lett.*, **10**, 4279 (2010).

89. A. I. Hochbaum, R. Chen, R. D. Delgado, W. Liang, E. C. Garnett, M. Najarian, A. Majumdar, and P. Yang, Enhanced thermoelectric performance of rough silicon nanowires. *Nature*, **451**, 163 (2008).

90. A. I. Boukai, Y. Bunimovich, J. Tahir-Kheli, J.-K. Yu, W. A. Goddard Iii, and J. R. Heath, Silicon nanowires as efficient thermoelectric materials. *Nature*, **451**, 168 (2008).

91. P. Martin, Z. Aksamija, E. Pop, and U. Ravaioli, Impact of phonon-surface roughness scattering on thermal conductivity of thin Si nanowires. *Phys. Rev. Lett.*, **102**, 125503 (2009).

92. Y. He and G. Galli, Microscopic origin of the reduced thermal conductivity of silicon nanowires. *Phys. Rev. Lett.*, **108**, 215901 (2012).

93. J. Lim, K. Hippalgaonkar, S. C. Andrews, A. Majumdar, and P. Yang, Quantifying surface roughness effects on phonon transport in silicon nanowires. *Nano Lett.*, **12**, 2475 (2012).

94. J. M. Ziman, *Electrons and Phonons: The Theory of Transport Phenomena in Solids,* Oxford University Press, UK, 1960.

95. M. C. Wingert, Z. C. Y. Chen, E. Dechaumphai, J. Moon, J.-H. Kim, J. Xiang, and R. Chen, Thermal conductivity of Ge and Ge–Si core–shell nanowires in the phonon confinement regime. *Nano Lett.*, **11**, 5507 (2011).

96. P. E. Hopkins, C. M. Reinke, M. F. Su, R. H. Olsson, E. A. Shaner, Z. C. Leseman, J. R. Serrano, L. M. Phinney, and I. El-Kady, Reduction in the thermal conductivity of single crystalline silicon by phononic crystal patterning. *Nano Lett.*, **11**, 107 (2010).

97. J.-K. Yu, S. Mitrovic, D. Tham, J. Varghese, and J. R. Heath, Reduction of thermal conductivity in phononic nanomesh structures. *Nat. Nanotechnol.*, **5**, 718 (2010).

98. M. N. Luckyanova, J. Garg, K. Esfarjani, A. Jandl, M. T. Bulsara, A. J. Schmidt, A. J. Minnich, S. Chen, M. S. Dresselhaus, Z. Ren, E. A. Fitzgerald, and G. Chen, Coherent phonon heat conduction in superlattices. *Science*, **338**, 936 (2012).

99. W. Kim, J. Zide, A. Gossard, D. Klenov, S. Stemmer, A. Shakouri, and A. Majumdar, Thermal conductivity reduction and thermoelectric figure of merit increase by embedding nanoparticles in crystalline semiconductors. *Phys. Rev. Lett.*, **96**, 045901 (2006).

100. G. Joshi, H. Lee, Y. Lan, X. Wang, G. Zhu, D. Wang, R. Gould, D. Cuff, M. Tang, M. Dresselhaus, G. Chen, and Z. Ren, Enhanced thermoelectric figure-of-merit in nanostructured p-type silicon germanium bulk alloys. *Nano Lett.*, **8**, 4670 (2008).

101. Y. Lan, B. Poudel, Y. Ma, D. Wang, M. Dresselhaus, G. Chen, and Z. Ren, Structure study of bulk nanograined thermoelectric bismuth antimony telluride. *Nano Lett.*, **9**, 1419 (2009).

102. E. F. Hockings, The thermal conductivity of silver antimony telluride. *J. Phys. Chem. Solids*, **10**, 341 (1959).

103. D. T. Morelli, V. Jovovic, and J. P. Heremans, Intrinsically minimal thermal conductivity in cubic I–V–VI$_2$ semiconductors. *Phys. Rev. Lett.*, **101**, 035901 (2008).

104. MaJ, DelaireO, A. F. May, C. E. Carlton, M. A. McGuire, L. H. VanBebber, D. L. Abernathy, EhlersG, T. Hong, HuqA, W. Tian, V. M. Keppens, Y. Shao Horn, and B. C. Sales, Glass-like phonon scattering from a spontaneous nanostructure in AgSbTe$_2$. *Nat. Nanotechnol.*, **8**, 445 (2013).

105. G. A. Slack, The thermal conductivity of nonmetallic crystals. *Solid State Phys.*, **34**, 1 (1979).

106. D. G. Cahill, S. K. Watson, and R. O. Pohl, Lower limit to the thermal conductivity of disordered crystals. *Phys. Rev. B*, **46**, 6131 (1992).

107. C. Chiritescu, D. Cahill, N. Nguyen, D. Johnson, A. Bodapati, P. Keblinski, and P. Zschack, Ultralow thermal conductivity in disordered, layered WSe$_2$ crystals. *Science*, **315**, 351–353,(2007).

Chapter 4

New Design Rules for Polymer-Based Thermoelectric Nanocomposites

Jeffrey J. Urban and Nelson E. Coates

4.1 Introduction

Thermoelectric materials convert thermal gradients into electrical gradients; this is a powerful concept in energy conversion which complements other approaches which rely on solar or mechanical inputs for direct or co-generation applications. Historically, thermoelectric materials have been composed of single phase narrow band gap semiconductors whose doping concentration was optimized to yield the best performance. Guidance on these materials was provided by tractable coupled transport equations based upon single parabolic bands and noninteracting electron approximations. Performance of these materials and innovations in materials design remained relatively stagnant for decades. In the early 1990s, however, the seminal papers of Hicks and Dresselhaus initiated an abundance of renewed interest in the field; this paper hypothesized that control over the dimensionality of a material could enable incredibly high performance in thermoelectrics.[1,2] These effects were fundamentally quantum in nature, and relied upon materials patterned at the nanoscale — where the fundamental length scales of charge and heat transport converge.

Ultimately, this powerful notion has led to dramatic improvements in thermoelectric performance in the most recent 20 years which far surpassed that of the preceding 40.[3] One common theme which crosscuts all these high-performance materials is the introduction of increased complexity. Complex thermoelectric materials consisting of mixtures of phases,[4] patterned superlattices,[5,6] embedded structures,[7] hierarchical structural elements,[8] and composite materials[9] are all different manifestations of this broad notion and comprise a rich and robust effort in the community today. In this chapter, we choose to focus on the newest of these areas–polymer nanocomposite thermoelectrics. These materials (typically) consist of two intertwined phases of materials — one of which is a polymer, and the other of which is either an inorganic or metallic nanomaterial, or a carbon-based nanomaterial (graphene, carbon nanotubes). While this is currently a nascent area and not yet mature, exciting developments highlighting new transport concepts, extraordinarily low thermal conductivity, and scalable processing are already attracting much attention. Here we review the basic physics of transport in these materials, how these concepts lead to important design rules, and demonstrate the future potential for polymer nanocomposites to be a breakthrough class of materials that open up new applications due to their flexibility and mechanical robustness.

The physical properties of composites can often be described as arithmetic averages of their components, that is, Maxwell–Garnett for dielectrics, Bruggeman for electrical conductivity, and Vegard's law for lattice parameters of alloys. For thermoelectrics, the Seebeck coefficient (S), electrical conductivity (σ), and thermal conductivity (κ) have generalized mixing rules, as elucidated by Levy[10] and Fel[11] in the 1990s.[12] These effective medium "averaging" composite theories have been used to accurately describe the properties of bulk composite materials, and predict that that highest achievable performance is limited by the best-performing individual component of the composite. However, in organic–inorganic nanoscale composites, there exist several routes for overcoming this limitation, leading to new collective behavior which can *surpass* those of either the individual components (see Fig. 4.1). This introduces new possibilities for the design of polymer–nanocrystal- and polymer–nanotube-based materials.

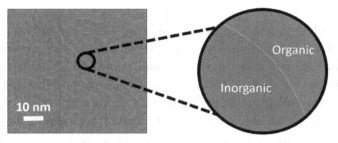

Figure 4.1. Cartoon representation of a nanoscale organic–inorganic composite with the large interfacial area.

In general, the breakdown of these simple composite "rules" generally occurs when the length scale of the physical phenomena corresponding to the observable is on the order of length scale of composite constituents. This breakdown of "averaging" gives rise to richer physical phenomena in many classes of functional materials, that is, ionic transport or dielectric tuning. In thermoelectric composites, the interaction of components also gives rise to richer physical phenomena, and provides many promising novel routes for improving their performance.

In this chapter, we want to explore these interactions in composites, and ask the question: "How would you try to tailor interactions in composites to form high-performance thermoelectric materials?" The design rules for doing so are just now being uncovered, and it is clear this is a rich field with many new opportunities.

4.2   Transport Background for Thermoelectrics

Historically, thermoelectrics have been evaluated based on their figure of merit, $ZT = \frac{S^2\sigma}{\kappa_e + \kappa_{ph}}$, where the numerator $S^2\sigma$ is referred to as the power factor, κ_e is the thermal conductivity due to the conduction of charges, and κ_{ph} is the thermal conductivity due to lattice vibrations. Arguably, ZT is not the best or most practical metric for thermoelectrics, with cost/watt, or even cost/area as potentially a more useful metric of comparison.[13–15] Factoring in these considerations, polymer-composite thermoelectrics have a promising scientific future.

Given that there is no existing theoretical framework for transport in polymer nanocomposites yet, we begin by introducing the canonical model for understanding thermoelectricity in solids. For band-conduction solids, this is the well-known Boltzmann transport equation for charges in the relaxation time approximation, which describes the evolution of particles with density and temperature in a volume of phase space that is perturbed from equilibrium with a force field (in this case, particle charge times electric field, $q\vec{E}$), and which takes the following form:

$$\frac{\partial f}{\partial t} + \frac{\vec{p}}{m^*} \cdot \nabla f + q\vec{E} \cdot \frac{\partial f}{\partial \vec{p}} = \frac{f_0 - f}{\tau},$$

where f is the distribution function that describes the number of particles per volume of phase space, $\vec{p}$ is the momentum, m^* is the effective mass, τ is the carrier relaxation time, and f_0 is the equilibrium Fermi–Dirac distribution defined as $f_0(E) = (e^{(E-E_F)/k_B T} + 1)^{-1}$ with E and E_F being energy and Fermi energy, T the temperature, and k_B the Boltzmann constant. In steady state, and assuming the deviation from equilibrium is small, the Boltzmann equation simplifies to

$$\frac{\vec{p}}{m^*} \cdot \nabla f_0 + \frac{\vec{p}}{m^*} \cdot q\vec{E}\frac{\partial f_0}{\partial E} = \frac{f_0 - f}{\tau}.$$

Now expanding for ∇f_0, we have

$$\frac{\vec{p}}{m^*} \cdot \left[-\nabla E_F - \frac{E - E_F}{T}\nabla T - q\nabla\phi \right] \frac{\partial f_0}{\partial E} = \frac{f_0 - f}{\tau},$$

with ϕ being the electrical potential. Defining $V = \frac{E_F}{q} + \phi_e$, we solve for our distribution function:

$$f = f_0 - \tau\frac{\vec{p}}{m^*} \cdot \left[-q\nabla V - \frac{E - E_F}{T}\nabla T \right] \frac{\partial f_0}{\partial E}.$$

The charge flux $\vec{J}$ resulting from the distribution function is given by

$$\vec{J} = \int_{E=0}^{\infty} fD(E)q\frac{\vec{p}}{m^*}dE,$$

with $D(E)$ as the density of states. Substituting for f:

$$\vec{J} = \int_{E=0}^{\infty} f_0 D(E) q \frac{\vec{p}}{m^*} dE$$

$$+ \int_{E=0}^{\infty} D(E) q \frac{\vec{p}}{m^*} \left(\tau \frac{\vec{p}}{m^*} \cdot \left[-q\nabla V - \frac{E - E_F}{T} \nabla T \right] \frac{\partial f_0}{\partial E} \right) dE.$$

For which the first term is zero, and together with the substitution $E = \frac{\vec{p} \cdot \vec{p}}{2m^*}$, the charge flux simplifies to

$$\vec{J} = \frac{2q^2}{3m^*} \int_{E=0}^{\infty} D(E) E \left(\tau \left[-\nabla V - \frac{E - E_F}{T} \nabla T \right] \frac{\partial f_0}{\partial E} \right) dE.$$

For the case where there is zero temperature gradient, and with the electrical conductivity σ defined from $\vec{J} = -\sigma \nabla V$, we obtain our final expression for the electrical conductivity in the relaxation time approximation:

$$\sigma = -\frac{2q^2}{3m^*} \int_{E=0}^{\infty} \frac{\partial f_0}{\partial E} D(E) E \tau(E) dE. \tag{4.1}$$

Next, we consider the energy flux $\vec{\Omega}$ for the nonequilibrium distribution function f, which is given by

$$\vec{\Omega} = \int_{E=0}^{\infty} f D(E) E \frac{\vec{p}}{m^*} dE$$

and expanding with our function f, we have

$$\vec{\Omega} = \int_{E=0}^{\infty} f_0 D(E) E \frac{\vec{p}}{m^*} dE$$

$$+ \int_{E=0}^{\infty} D(E) E \frac{\vec{p}}{m^*} \left(\tau \frac{\vec{p}}{m^*} \cdot \left[-q\nabla V - \frac{E - E_F}{T} \nabla T \right] \frac{\partial f_0}{\partial E} \right) dE.$$

Again for which the first term is zero, and together with the substitution $E = \frac{\vec{p} \cdot \vec{p}}{2m^*}$, simplifies to

$$\vec{\Omega} = \frac{2}{3m^*} \int\limits_{E=0}^{\infty} D(E)E^2 \left(\tau \left[-\nabla V - \frac{E - E_F}{T} \nabla T \right] \frac{\partial f_0}{\partial E} \right) dE.$$

The Seebeck coefficient S is defined by the equation: $\nabla V = -S\nabla T$, and solving for the nonzero temperature contribution to S with the energy flux equation, we find the Boltzmann formulation of the Seebeck coefficient with ν^2 as the square of the velocity:

$$S = \frac{1}{qT} \left(\frac{\int\limits_{E=0}^{\infty} (E - E_F)\frac{\partial f_0}{\partial E} D(E)\nu^2\tau(E)dE}{\int\limits_{E=0}^{\infty} \frac{\partial f_0}{\partial E} D(E)\nu^2\tau(E)dE} \right). \tag{4.2}$$

Additionally, in the adiabatic scattering approximation, one can show that these expressions can be written as the well-known Mott formulations, with n being the carrier distribution, and μ being the carrier mobility:[16]

$$\sigma = n(E)q\mu(E), \tag{4.3}$$

$$S = \frac{\pi^2 k_B^2 T}{3q} \left[\frac{1}{n}\frac{dn(E)}{dE} + \frac{1}{\mu}\frac{d\mu(E)}{dE} \right]_{E=E_F}. \tag{4.4}$$

The ratio of the electronic component of the thermal conductivity (k_e) to the electrical conductivity can be written as (known as the Wiedemann–Franz law):

$$\frac{k_e}{\sigma} = \frac{\pi^2 k_B^2 T}{3q^2}$$

and the phonon component of the thermal conductivity with C_{ph} being the heat capacity per unit volume, v_{ph} being the phonon velocity, and τ_{ph} being the phonon scattering time is given by

$$k_{ph} = \frac{1}{3}C_{ph}v_{ph}^2\tau_{ph}.$$

Generally, it can be seen from these equations that S and σ are anticorrelated; as n or μ increases, S decreases. This fundamental relationship has been one of the main reasons that the power factor $(S^2\sigma)$ in bulk thermoelectric materials had seen so little improvement since the advances reached in the 1950s. However, there are many fundamentally new routes for optimization in nanoscale composites — it is indeed possible to increase $\frac{dn(E)}{dE}$ or $\frac{d\mu(E)}{dE}$ without decreasing n or μ, or where S is not substantially sensitive to either n or μ, and thus σ can be increased without significantly decreasing S. This is one of the many novel mechanisms by which composite thermoelectric materials provide new avenues for improvement.

In this chapter, we hope to elucidate several of these routes for optimizing organic–inorganic materials, and present relevant examples and evidence for composites where these routes are being realized.

4.3 Carrier Filtering and Energy Level Alignment: Engineering $\mu(E)$, and $\tau(E)$

One of the most promising routes for increasing thermoelectric performance in organic–inorganic composites can be achieved by engineering potential barriers across the organic–inorganic junction. Carriers with energies lower than the potential barrier are scattered, leading to a decrease in the electrical conductivity, but an increase in the Seebeck coefficient and power factor of the composite material. This basic idea of energy-selecting transport was first demonstrated in inorganic superlattices,[17,18] but for organic–inorganic composites, was first demonstrated in single-molecule junctions, with the theory described by Paulson and Datta in 2003,[19] followed by an experimental report by Reddy *et al.* in 2007.[20] Often termed "carrier filtering" or "energy filtering" (see Fig. 4.2) the basic concept is to increase the power factor of the system by selecting for carriers whose energy offset from the Fermi level $(E - E_F)$ is large. As we derived in the previous section (Eq. (2)), this energy offset is directly proportional to the Seebeck coefficient. The enhancement in thermoelectric properties arises fundamentally from a sharp change

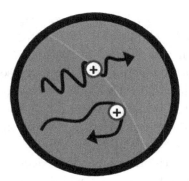

Figure 4.2. Cartoon representation of energy filtering at an organic–inorganic junction. High-energy carriers traverse the junction, while low-energy carriers are scattered.

in the energy dependence of the relaxation time, and thus the carrier mobility, $\mu(E)$, (or transmission, in the single-molecule case) as carriers move between the organic and inorganic materials. Although the entropy per carrier is increased through the sharp energy-dependent change in scattering across the organic–inorganic junction, one is decreasing the overall number of carriers, so the electrical conductivity is thereby decreased (Fig. 4.3). To fully capitalize on such a peaked scattering mechanism across the junction, the carrier transport (transmission channels in the single molecule case, $\mu(E)$ in the bulk composite case) in either the organic or inorganic materials should have a sharp contrast (i.e., discrete energy states, such as those found in nanomaterials or organic molecules, are best), and the Fermi level of the composite should rest within a few $k_B T$ of the point where the power factor $S^2\sigma$ is maximized. Although the relaxation time approximation that forms the basis of Boltzmann transport theory is not in general valid for a highly inhomogeneous material with many scattering times, it is nevertheless instructive to consider the example of transport across a potential barrier of height E_b, when the value of the chemical potential of the composite system is also E_b. Classically, carriers with energy $E < E_b$ have a relaxation time of zero, and carriers with an energy $E > E_b$ have a relaxation time of τ_0. In order to understand how these scattering times change the thermoelectric properties, first let us examine the

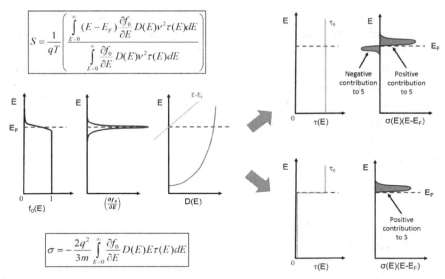

Figure 4.3. Cartoon diagrams of (in order) the Fermi distribution function, derivative of the Fermi distribution function, density of states, and (upper diagram) Seebeck coefficient contributions without any potential barrier, and (lower diagram) Seebeck coefficient through a potential barrier at $\tilde{\mu}$.

Boltzmann transport expression for the electrical conductivity:

$$\sigma = -\frac{2q^2}{3m} \int\limits_{E=0}^{\infty} \frac{\partial f_0}{\partial E} D(E) E \tau(E) dE.$$

Assuming a roughly constant value for $D(E)$ (such as the case when E_F and E_b are deep in a band), when we substitute our Heaviside function for $\tau(E)$ and integrate over $\frac{\partial f_0}{\partial E} D(E) E$, our electrical conductivity is reduced by a factor of 2. Next, let us examine the Seebeck coefficient:

$$S = \frac{1}{qT} \left(\frac{\int\limits_{E=0}^{\infty} (E - E_F) \frac{\partial f_0}{\partial E} D(E) v^2 \tau(E) dE}{\int\limits_{E=0}^{\infty} \frac{\partial f_0}{\partial E} D(E) v^2 \tau(E) dE} \right)$$

again, looking at the condition where for $E < E_b$, $\tau(E) = 0$, and for $E > E_b$, $\tau(E) = \tau_0$. The integral in the numerator which would have been close to zero for a constant τ, is now maximized, and the integral in the denominator is roughly half its previous value. Therefore, the Seebeck coefficient can be orders of magnitude larger, (infinitely larger assuming a constant $D(E)$) and thus the thermoelectric power factor will be greatly increased.

Carrier filtering has been demonstrated multiple times in single-molecule junctions[21–24] as well as in some all-inorganic systems such as core–shell nanoparticle films,[25] superlattices,[26] and through grain boundaries.[27] However, reports of carrier filtering in bulk organic–inorganic composites remain scant, and subject to debate.[28]

4.4 Engineering Power Factor Enhancement through "Trap" States: $D(E)$, $n(E)$

Carrier traps in organic–inorganic composites represent the movement of a higher energy electronic state to a lower energy one, and are often thought to arise from changes in bonding along the surface of the inorganic material, but can also result from changes in bonding in the organic material. The role of these interfacial traps in the charge transport of composites has been studied in devices such as field-effect transistors, photovoltaics, and photodetectors, but their potential for improving the thermoelectric performance of composite materials has yet to be realized, and is the subject of this section.

Conceptually, trap engineering in organic–inorganic composites is similar to density of states engineering in inorganic alloys and composites, where distortions in a parabolic density of states can be induced through the introduction of impurities with defined energy levels. However, given that these impurities can only really be atomic species that can alloy to some degree, or sit within the host lattice, the number of options are quite limited. By contrast, in the organic–inorganic composite case, the "impurities" could be changes in bonding in the organic material, or on the surface of the inorganic material, or structural distortions in the organic material, or some combination of electronic and morpohological modifications. In order to improve the thermoelectric properties of the composite, the result of the modification should be a high-density of carrier traps close in energy to the transport band edge, (or E_F) of the material. To see how this trap density improves the thermoelectric performance, we refer to the Mott formulation of the Seebeck coefficient:

$$S = \frac{\pi^2 k_B^2 T}{3q} \left[\frac{1}{n} \frac{dn(E)}{dE} + \frac{1}{\mu} \frac{d\mu(E)}{dE} \right]_{E=E_F}.$$

From this equation, it can be seen that a change in $\frac{dn(E)}{dE}$ without a significant increase in n near E_F results in an increase in S. A sharply defined density of carrier traps could have a large effect on $\frac{dn(E)}{dE}$, and because the trap resulted from the change in occupation of a higher level electronic state to the lower energy state, n can be effectively decoupled from its derivative, and σ can remain largely unaffected. A cartoon of an engineered trap density of states that could result in high thermoelectric performance is shown in Fig. 4.4.

There are several examples of conceptually similar density of states engineering in purely inorganic alloys and composites. Heremans *et al.* report the increase in Seebeck coefficient in PbTe through introduction of Thallium impurities which increase the density of states near the Fermi level.[29] Similarly, Paul *et al.* enhance the Seebeck coefficient in PbTe:Cr through introduction of iodine impurities, which also establishes a new band of states close to the

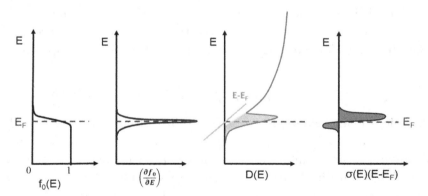

Figure 4.4. Cartoon diagrams of (in order) the Fermi distribution function, derivative of the Fermi distribution function, density of states including the shaded trap density, and increased Seebeck coefficient resulting from the trap density-of-states.

Fermi level.[30] The formation of a mid-gap band of states (likely due to changes in surface bonding) has been demonstrated in semiconductor nanocrystals films, raising the possibility of engineering placement of that band of states using interfacial bonding of organic–inorganic materials.[31] In the new organic–inorganic perovskites, the formation of a mid-gap defect band has also been demonstrated using oxygen vacancies, with the Fermi level located in that band. Metal-like conductivities are observed in these samples, and presumably the room temperature Seebeck coefficient is larger as well.[32] In purely organic electronic materials, changes in $D(E)$ have been demonstrated that simultaneously improve both S and σ through the reduction of disorder-induced trap states.[33] Similarly, gating Pentacene and Rubreneto move the Fermi energy and fill trap states near the band edge has been shown to manipulate the thermoelectric properties, and can result in a simultaneous increase in both S and σ.[34,35]

Clearly, there is a significant potential to improve thermoelectric performance by tuning $D(E)$ through the defined introduction of trap densities in inorganic–organic composites. In composite materials, trap state engineering could be accomplished through changes in bonding in the organic or inorganic material, or through structural disorder in the materials, through some combination of the two. To our knowledge, no demonstration of trap energy level engineering exists in organic–inorganic composite thermoelectric literature, and thus this route for materials improvement is ripe for development.

4.5 Heterojunction Charge Transfer: n, μ, and $\frac{d\mu(E)}{dE}$

Organic and inorganic electronic materials often have significantly different carrier densities, energy levels, and densities of states. Therefore, when combining these materials together in a composite, charges are likely to move across the organic–inorganic heterojunction to reach equilibrium (Fig. 4.5). Given the ability to tune carrier densities through doping in inorganic materials, as well as the wide range of chemical donor or accepter moieties to choose from when designing organic electronic materials, a rich diversity

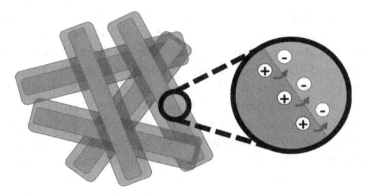

Figure 4.5. Cartoon representation of charge transfer at an organic–inorganic junction. The movement of charges across the interface can improve thermoelectric properties by reducing the coupling between mobility and carrier density, or by tuning carrier concentrations to increase the thermoelectric power factor of the composite.

of heterojunction charge transfer behavior can be engineered and to enhanced thermoelectric transport. Due to the physical considerations outlined here, one will generally have some sort of charge transfer between closely coupled organic and inorganic components in these systems — the challenge is to productively direct the flow of these carriers, either changing carrier densities, scattering mechanisms or filling trap states to improve the overall performance of these unique materials.

One of the ways in which heterojunction charge transfer can improve the thermoelectric properties of composites is by spatially separating charges from their charge donating constituents, which reduces impurity scattering (often a dominant scattering mechanism in inorganic materials) and coupling between n and μ in the composite. This phenomenon is, in classical thin-film semiconductor literature, known as "modulation doping" typically relies on carefully grown MOCVD films; it is exciting to realize these same physics in nanocomposites. Recently, Zebarjadi *et al.*[36] has accomplished modulation doping for both p-type and n-type materials using alloys of silicon and germanium. Here, energy band engineering in a pressed inorganic composite was used to drive charge transfer from nanoparticles into a host matrix, effectively using nanoparticles as "dopants." The engineered heterojunction structure resulted in spatial separation of the ionized dopant atoms, which reduced the ionized impurity scattering of the charge transfer carriers, and thus led to larger mobility and thermoelectric power factor in the composite films. We note that because there is no state mixing, modeling of the properties of the composite film here followed effective medium approximations, except with an increased carrier density and no concomitant reduction in mobility of the host material. Other approaches leveraging heterojunction charge transfer abound in the large field of composites formed from single-walled carbon nanotubes mixed with conjugated polymers; there exist several examples of these materials where electron transfer from the polymer to nanotube is posited to result in an increase in power factor either due to the high-mobility carrier transport along the axis of the nanotube, or doping of the conjugated polymer.[37-39] Heterojunction charge transfer effects may also occur in high-performing thermoelectric composites comprised of Tellurium nanowires and poly(3,4-ethylenedioxythiophene): poly(styrenesulfonate) (PEDOT:PSS), where the composite exhibits a higher electrical conductivity and power factor than either of its constituents.[40-42] The authors also note other possible morphological and electronic structure hybridization effects here as well.

Another way in which charge transfer between the organic and inorganic components of a composite can improve thermoelectric properties is by *reducing* the carrier concentration to optimize the Seebeck coefficient of one of the components, particularly if one component has a large (metallic-like) carrier density. From the Mott relation

$$S = \frac{\pi^2 k_B^2 T}{3q} \left[\frac{1}{n} \frac{dn(E)}{dE} + \frac{1}{\mu} \frac{d\mu(E)}{dE} \right]_{E=E_F}$$

and so for a material whose Seebeck coefficient is limited by a large number of carriers, n, driving charge transfer out of that material can increase the thermoelectric power factor. Such a mechanism of increasing the Seebeck coefficient has been suggested in composite films of PEDOT:PSS and gold nanoparticles, where hole transfer from the PEDOT:PSS to the gold nanoparticles may be responsible for the observed increase in Seebeck coefficient of the composite.[43]

As presented in this section, heterojunction charge transfer is fundamentally ubiquitous in organic–inorganic compound materials, and presents powerful opportunities to improve their thermoelectric performance through either modulation doping, or tuning carrier densities to optimize the performance of the composite. Although there is insufficient evidence at present in the literature for the following mechanisms, heterojunction charge transfer also has the potential to improve the thermoelectric performance of inorganic–organic composites by tuning energy levels through nanoscale size control of inorganic materials, and by chemically controlling energy levels in organic materials. These approaches have yet to be developed, but are promising and worth exploring.

4.6 Reducing Kappa: $\kappa = \kappa_e + \kappa_{ph}$

4.6.1 Carrier Contributions to Thermal Conductivity: κ_e

In the introduction of this chapter, we presented the Wiedemann–Franz law, which describes the relationship between the thermal and electrical conductivities for a three-dimensional free electron gas: $\frac{k_e}{\sigma} = \frac{\pi^2 k_B^2 T}{3q^2} = LT$ with L as the Lorenz number. The Wiedemann–Franz law states that as the electrical conductivity of a material increases, so does its thermal conductivity. One can see therefore from substitution into the equation $ZT = \frac{S^2\sigma}{\kappa_e+\kappa_{ph}} = \frac{S^2\sigma}{\sigma LT+\kappa_{ph}}$, that material efficiency becomes limited in the limit of high electrical conductivity. The Wiedemann–Franz law is widely cited as a fundamental barrier to improving thermoelectric efficiency, and a large variety of metals or degenerately doped semiconductors are hampered by this "law." It would therefore be exciting to alter the relationship between κ_e and σ. Indeed, it has been shown that there exist transport regimes in which there is no linear relationship between k_e and σ, and in fact k_e and σ can become entirely decoupled in 1D systems.[44,45] As an example, we start with low-dimensional conductors and follow Jeong *et al.*[46] for the limiting case of a single conduction channel (perfectly 1D system), where $D(E) = \delta(E - E_0)$. For this density of states relationship, the electrical conductivity reduces to: $\sigma = -\frac{2q^2}{3m^*} \left.\frac{\partial f_0}{\partial E}\right|_{E_0} E_0 \tau(E_0) \equiv \sigma_0$, and the Seebeck coefficient becomes $S = \frac{(E_0 - E_F)}{qT}$. The electronic heat conduction is given by $k_e = k_0 - S^2 \sigma_0 T$, and for a single channel with zero potential gradient, k_0 is given by $k_0 = -\frac{2q^2 T}{3m^*} \left.\frac{\partial f_0}{\partial E}\right|_{E_0} E_0 \tau(E_0) \left(\frac{(E_0 - E_F)}{qT}\right)^2$. Therefore, for the single conduction channel case, $k_e = k_0 - S^2 \sigma_0 T = 0$ and thus $\frac{k_e}{\sigma} = 0$, which violates the Wiedemann–Franz law in the most extreme (and beneficial) way possible.

Achieving 1D conduction in real physical systems may be possible in both organic and inorganic systems. The conducting bands in many organic systems can be made quite sharp, similar to a delta function. This sharpness in $D(E)$ is the result of quasi-1D crystallinity, stemming from either elongated linear conjugated polymer chains, or through a pi-stacking crystallization of small molecules. With analogy to the single conducting channel case, the ratio of $\frac{k_e}{\sigma}$ in some ideal organic systems is predicted to be quite small, resulting in a very high thermoelectric performance.[47,48] Although these predictions of high performance have not been experimentally confirmed, template bonding or other physical interactions between organic and inorganic materials in hybrid composites may be a promising route for realizing high-performance quasi 1D electronic transport structures.

4.6.2 Lattice Contributions to Thermal Conductivity: κ_{ph}

One of the major approaches for improving the thermoelectric performance of materials in the last decade has been through a reduction thermal conductivity effected by nanostructuring materials to inhibit phonon transport; this has been reviewed extensively elsewhere.[49,50] For organic–inorganic composites, the phonon component of the thermal conductivity is intrinsically low due to the acoustic impedance and mismatch between vibrational densities of state at the organic–inorganic interface. Additionally, organic–inorganic nanocomposites often have a large number of arbitrarily distributed surfaces and boundaries and nanoscale structures of various sizes and shapes that also efficiently scatter phonons and further reduce the thermal conductivity of the composite.[51] As a result, it is a general finding that organic–inorganic composite materials of many types exhibit very low thermal conductivities — the hard work in engineering optimal thermoelectric performance in these materials is often in the power factor. Despite these overall favorable values, many good scientific questions remain which can be of further value. For example, until recently, interfacial thermal impedance had been difficult to directly measure, except in self-assembled monolayer geometries[52] but had been proposed as an explanation for observed lower thermal conductivities than predicted from simplistic noninteracting composite models.[53] A recent report from Ong *et al.*[54] describes thermal conductivity measurements for arrays of inorganic nanocrystals coated with organic ligands, and finds that the thermal conductivity of the composite depends strongly on the chemistry and density of the organic–inorganic interfaces, as well as on the volume fractions of the constituents, and depends only weakly on the thermal conductivity of the inorganic core. This result suggests that engineering the chemistry and controlling the alignment of vibrational state densities in an organic–inorganic composite could be used to tailor the temperature drops across individual components of a composite, and also significantly reduce (or increase) the composite thermal conductivity. In addition to interface thermal resistances, an additional mechanism for reducing the phonon component of the thermal conductivity in organic–inorganic composites can be through interfacial interactions that change the vibrational density of states, and thus thermal conductivity of an individual component. This mechanism for the reduction in thermal conductivity of a composite below effective medium theories has been suggested in polymer–carbon nanotube composites where both the thermal boundary resistance and a reduction in the thermal conductivity of the carbon nanotube suppress the thermal conductivity of the composite.[55] It is clear that as increasing control over material morphology and composition is achieved, many very

fundamental questions about how bonding modes across hard–soft
interfaces contribute to the transmission of thermal energy remain
to be answered.

4.7 Structural Changes in the Organic Material at the Organic–Inorganic Interface: $S(S_1, S_2, S_{\text{interface}})$, $\sigma(\sigma_1, \sigma_2, \sigma_{\text{interface}})$

Transport in organic electronic materials is very sensitive to
morphology. Due to confinement (i.e., physical constraints) or bonding, the morphology of the organic at the organic–inorganic interface can be changed. These morphological changes often translate
into modified transport properties and, for example, can be used
to increase the carrier mobility, which increases σ in proportion
to the volume of the interfacial region. From the Mott relation,
$S = \frac{\pi^2 k_B^2 T}{3q} \left[\frac{1}{n} \frac{dn(E)}{dE} + \frac{1}{\mu} \frac{d\mu(E)}{dE} \right]_{E=E_F}$ and thus it can be seen that if
the value of the Seebeck coefficient of the organic component is dom-
inated by $\frac{1}{n} \frac{dn(E)}{dE}$ and not $\frac{1}{\mu} \frac{d\mu(E)}{dE}$ such as the case when a material is
near the metal–insulator transition[33,56] or if the Seebeck coefficient
of the composite is largely dominated by the inorganic phase of the
material and the transport in the composite is a series-connected
one,[10,11] the power factor of the composite can be improved considerably by an increase in mobility in the interfacial organic material.
A phenomenological model that includes an interfacial volume component of organic material with enhanced thermoelectric properties
has been used to describe the behavior of an organic–inorganic composite composed of Tellurium nanowires and PEDOT:PSS. Although
the authors are agnostic about the mechanism by which the thermoelectric properties of the interfacial volume are changed (i.e., charge
transfer at the interface, structural changes at the interface, etc.)
the thermoelectric behavior of the composite could not be explained
with standard mixing models, and this interfacial-phase model provides a framework for considering changes in the organic material
at the organic–inorganic interface.[41] There are several additional
sets of data in the organic–inorganic composite literature that could
likely be modeled with an additional interfacial organic phase with

an increased electrical conductivity.[53,57] However, due to the difficulty in characterizing a small structural change in a small volume of organic material in a composite, conclusive evidence for such a mechanism is sparse. Nevertheless, the importance of control over interfacial and junction resistance is further highlighted by examples where the interfacial resistance is large, and causes a significantly hampered carrier transport.[58]

The correspondence between morphology and transport can be leveraged in other ways as well. As discussed earlier in this chapter, another potential route for improving thermoelectric performance in composites using structural changes in the organic at the organic–inorganic interface is through control over dimensionality of carrier transport. This is an exciting possibility, and presents a promising route for sharpening $D(E)$ and thus $n(E)$, and also for suppressing the ratio of $\frac{k_3}{\sigma}$.

4.8 Summary, Outlook, and Conclusion

In this chapter, we have described the fundamental rules of design for optimizing thermoelectric transport in polymer-composite and hybrid materials. By doing so, we endeavor to show that these materials present fundamentally new levers for transport, are amenable to rational engineering and design, and deserve to be considered alongside more traditional thermoelectric materials. The sharp contrasts in transport mechanism, vibrational modes, morphology, and bonding energy between organic and inorganic materials provides much wide-open territory for researchers to craft "custom-built" materials. While still a nascent field, there remain many attractive scientific targets in this area due to the extensive complexity of these materials. For instance, there is no simple long-range periodic potential, thus, even something as basic as developing an atomistic portrait for how energy transport occurs in these materials is incredibly challenging — these materials have complex structural and electronic behaviors that span fundamental length-scales ranging from the quantum mechanical to the classical. Indeed, as many reports now suggest strong renormalizing interactions between the polymer and inorganic components of composites, it is not excessive to view these as a new form of condensed matter, and one with exciting and untapped possibilities for a vast community of chemists, engineers, materials scientists, and physicists to explore.

References

1. L. D. Hicks and M. S. Dresselhaus, Effect of quantum-well structures on the thermoelectric figure of merit. *Phys. Rev. B*, **47**, 12727 (1993).
2. L. D. Hicks, and M. S. Dresselhaus, Thermoelectric figure of merit of a one-dimensional conductor, *Phys. Rev. B*, **47**, 16631 (1993).
3. C. J. Vineis, A. Shakouri, A. Majumdar, and M. G. Kanatzidis, Nanostructured thermoelectrics: Big efficiency gains from small features. *Adv. Mater.*, **22**, 3970 (2010).
4. J. Androulakis, C.-H. Lin, H.-J. Kong, C. Uher, C.-I. Wu, T. Hogan, B. A. Cook, T. Caillat, K. M. Paraskevopoulos, and M. G. Kanatzidis, Spinodal decomposition and nucleation and growth as a means to bulk nanostructured thermoelectrics: Enhanced performance in $Pb_{1-x}Sn_xTe-PbS$. *J. Am. Chem. Soc.*, **129**, 9780 (2007).
5. R. Venkatasubramanian, E. Siivola, T. Colpitts, and B. O'Quinn, Thin-film thermoelectric devices with high room-temperature figures of merit. *Nature*, **413**, 597 (2001).
6. T. C. Harman, P. J. Taylor, M. P. Walsh, and B. E. LaForge, Quantum dot superlattice thermoelectric materials and devices, *Science*, **297**, 2229 (2002).
7. K. F. Hsu, S. Loo, F. Guo, W. Chen, J. S. Dyck, C. Uher, T. Hogan, E. K. Polychroniadis, and M. G. Kanatzidis, Cubic $AgPbmSbTe_{2+m}$: Bulk thermoelectric materials with high figure of merit. *Science*, **303**, 818 (2004).
8. W. K. Liebmann and E. A. Miller, Preparation, phase-boundary energies, and thermoelectric properties of InSb–Sb eutectic alloys with ordered microstructures, *J. Appl. Phys.*, **34**, 2653 (1963).
9. A. R. Abramson, K. Woo Chul, S. T. Huxtable, Y. Haoquan, W. Yiying, A. Majumdar, T. Chang-Lin, and Y. Peidong, Fabrication and characterization of a nanowire/polymer-based nanocomposite for a prototype thermoelectric device. *J. Microelectromech. Syst.*, **13**, 505 (2004).
10. D. J. Bergman, and O. Levy, Thermoelectric properties of a composite medium, *J. Appl. Phys.*, **70**, 6821 (1991).
11. D. J. Bergman, and L. G. Fel, Enhancement of thermoelectric power factor in composite thermoelectric. *J. Appl. Phys.*, **85**, 8205 (1999).

12. A. Shakouri, Recent developments in semiconductor thermoelectric physics and materials, *Annu. Rev. Mater. Res.*, **41**, 399 (2011).
13. K. Yazawa, and A. Shakouri, Cost-efficiency trade-off and the design of thermoelectric power generators. *Environ. Sci. Technol.*, **45**, 7548 (2011).
14. S. K. Yee, S. LeBlanc, K. E. Goodson, and C. Dames, $ perW metrics for thermoelectric power generation: Beyond ZT. *Ener. Environ. Sci.*, **6**, 2561 (2013).
15. S. LeBlanc, S. K. Yee, M. L. Scullin, C. Dames, and K. E. Goodson, Material and manufacturing cost considerations for thermoelectric. *Renew Sust. Energ. Rev.*, **32**, 313 (2014).
16. M. Jonson, and G. D. Mahan, Mott's formula for the thermopower and the Wiedemann–Franz law. *Phys. Rev. B*, **21**, 4223 (1980).
17. D. Vashaee, and A. Shakouri, Electronic and thermoelectric transport in semiconductor and metallic superlattices. *J. Appl. Phys.*, **95**, 1233 (2004).
18. J. M. Zide, D. O. Klenov, S. Stemmer, A. C. Gossard, G. Zeng, J. E. Bowers, D. Vashaee, and A. Shakouri, Thermoelectric power factor in semiconductors with buried epitaxial semimetallic nanoparticles. *Appl. Phys. Lett.*, **87**, 112102 (2005).
19. M. Paulsson, and S. Datta, Thermoelectric effect in molecular electronics. *Phys. Rev. B*, **67**, 241403 (2003).
20. P. Reddy, S. Y. Jang, R. A. Segalman, and A. Majumdar, Thermoelectricity in molecular junctions. *Science*, **315**, 1568 (2007).
21. K. Baheti, J. A. Malen, P. Doak, P. Reddy, S. Y. Jang, T. D. Tilley, A. Majumdar, and R. A. Segalman, Probing the chemistry of molecular hetero junctions using thermoelectricity. *Nano Lett.*, **8**, 715 (2008).
22. J. A. Malen, P. Doak, K. Baheti, T. D. Tilley, R. A. Segalman, and A. Majumdar, Identifying the length dependence of orbital alignment and contact coupling in molecular hetero junctions, *Nano Lett.*, **9**, 1164 (2009).
23. J. A. Malen, S. K. Yee, A. Majumdar, and R. A. Segalman, Fundamentals of energy transport, energy conversion, and thermal properties inorganic–inorganicheterojunctions. *Chem. Phys. Lett.*, **491**, 109 (2010).
24. J. R. Widawsky, P. Darancet, J. B. Neaton, and L. Venkataraman, Simultaneous determination of conductance and thermopower of single molecule junctions. *Nano Lett.*, **12**, 354 (2011).
25. J.-H. Bahk, P. Santhanam, Z. Bian, R. Ram, and A. Shakouri, Resonant carrier scattering by core–shell nanoparticles for thermoelectric power factor enhancement. *Appl. Phys. Lett.*, **100**, 012102 (2012).

26. J. M. O. Zide, D. Vashaee, Z. X. Bian, G. Zeng, J. E. Bowers, A. Shakouri, and A. C. Gossard, Demonstration of electron filtering to increase the Seebeck coefficient in In(0.53)Ga(0.47)As/In(0.53)Ga(0.28)Al(0.19)As superlattices. *Phys. Rev. B*, **74**, 205335 (2006).

27. B. Moyzhes, and V. Nemchinsky, Thermoelectric figure of merit of metal–semiconductor barrier structure based on energy relaxation length. *Appl. Phys. Lett.*, **73**, 1895 (1998).

28. M. He, J. Ge, Z. Q. Lin, X. H. Feng, X. W. Wang, H. B. Lu, Y. L. Yang, and F. Qiu, Thermopower enhancement in conducting polymer nanocomposites via carrier energy scattering at the organic–inorganic semiconductor interface. *Ener. Environ. Sci.*, **5**, 8351 (2012).

29. J. P. Heremans, V. Jovovic, E. S. Toberer, A. Saramat, K. Kurosaki, A. Charoenphakdee, S. Yamanaka, and G. J. Snyder, Enhancement of thermoelectric efficiency in PbTe by distortion of the electronic density of states. *Science*, **321**, 554 (2008).

30. B. Paul, P. K. Rawat, and P. Banerji, Dramatic enhancement of thermoelectric power factor in PbTe:Crco-doped with iodine. *Appl. Phys. Lett.*, **98**, 262101 (2011).

31. P. Nagpaland V. I. Klimov, Role of mid-gap states in charge transport and photoconductivity in semiconductor nanocrystal films. *Nat. Commun.*, **2**, 486 (2011).

32. J. Ravichandran, W. Siemons, H. Heijmerikx, M. Huijben, A. Majumdar, and R. Ramesh, Anepitaxial transparent conducting perovskiteoxide: Double-doped SrTiO$_3$. *Chem. Mater.*, **22**, 3983 (2010).

33. O. Bubnova, Z. U. Khan, H. Wang, S. Braun, D. R. Evans, M. Fabretto, P. Hojati-Talemi, D. Dagnelund, J. B. Arlin, Y. H. Geerts, S. Desbief, D. W. Breiby, J. W. Andreasen, R. Lazzaroni, W. M. M. Chen, I. Zozoulenko, M. Fahlman, P. J. Murphy, M. Berggren, and X. Crispin, Semi-metallic polymers. *Nat. Mater.*, **13**, 190 (2014).

34. K. P. Pernstich, B. Rossner, and B. Batlogg, Field-effect-modulated Seebeck coefficient in organic semiconductors, *Nat. Mater.*, **7**, 321 (2008).

35. W. C. Germs, K. Guo, R. A. J. Janssen, and M. Kemerink, Unusual thermoelectric behavior indicating a hopping to band like transport transition in pentacene. *Phys. Rev. Lett.*, **109**, 016601 (2012).

36. M. Zebarjadi, G. Joshi, G. H. Zhu, B. Yu, A. Minnich, Y. C. Lan, X. W. Wang, M. Dresselhaus, Z. F. Ren, and G. Chen, Power factor enhancement by modulation doping in bulk nanocomposites. *Nano Lett.*, **11**, 2225 (2011).

37. H. Zengin, W. S. Zhou, J. Y. Jin, R. Czerw, D. W. Smith, L. Echegoyen, D. L. Carroll, S. H. Foulger, and J. Ballato, Carbon nanotube doped polyaniline, *Adv. Mater.*, **14**, 1480 (2002).

38. Y. F. Ma, W. Cheung, D. G. Wei, A. Bogozi, P. L. Chiu, L. Wang, F. Pontoriero, R. Mendelsohn, and H. X. He, Improved conductivity of carbon nanotube networks by *in situ* polymerization of a thin skin of conducting polymer. *Acs. Nano.*, **2**, 1197 (2008).

39. Q. Yao, L. D. Chen, W. Q. Zhang, S. C. Liufu, and X. H. Chen, Enhanced thermoelectric performance of single-walled carbon nanotubes/polyaniline hybrid nanocomposites. *Acs. Nano.*, **4**, 2445 (2010).

40. K. C. See, J. P. Feser, C. E. Chen, A. Majumdar, J. J. Urban, and R. A. Segalman, Water-processable polymer-nanocrystal hybrids for thermoelectrics. *Nano Lett.*, **10**, 4664 (2010).

41. N. E. Coates, S. K. Yee, B. McCulloch, K. C. See, A. Majumdar, R. A. Segalman, and J. J. Urban, Effect of interfacial properties on polymer-nanocrystal thermoelectric transport, *Adv. Mater.*, **25**, 1629 (2013).

42. J. N. Heyman, B. A. Alebachew, Z. S. Kaminski, M. D. Nguyen, N. E. Coates, and J. J. Urban, Terahertz and infrared transmission of an organic/inorganic hybrid thermoelectric material, *Appl. Phys. Lett.*, **104**, 141912 (2014).

43. N. Toshima, and N. Jiravanichanun, Improvement of thermoelectric properties of PEDOT/PSS films by addition of gold nanoparticles: Enhancement of Seebeck coefficient. *J. Electron. Mater.*, **42**, 1882 (2013).

44. A. Smontara, K. Biljakovic, and S. N. Artemenko, Contribution of charge-density-wavephase excitations to thermal-conductivity below the peierls transition. *Phys. Rev. B*, **48**, 4329 (1993).

45. C. L. Kane, and M. P. A. Fisher, Thermal transport in a Luttinger liquid. *Phys. Rev. Lett.*, **76**, 3192 (1996).

46. C. Jeong, R. Kim, and M. S. Lundstrom, On the best band structure for thermoelectric performance: A Landauer perspective. *J. Appl. Phys.*, **111**, 113707 (2012).

47. A. Casian, Violation of the Wiedemann–Franz law in quasi-one-dimensional organic crystals. *Phys. Rev. B*, **81**, 155415 (2010).

48. A. Casian, and V. Dusciac, Effect of a Lorenz pound number decrease on a thermoelectric pound efficiency in a quasi-one-dimensional pound organic crystals. *J. Electron. Mater.*, **42**, 2151 (2013).

49. M. S. Dresselhaus, G. Chen, M. Y. Tang, R. G. Yang, H. Lee, D. Z. Wang, Z. F. Ren, J. P. Fleurial, and P. Gogna, New directions for low-dimensional thermoelectric materials. *Adv. Mater.*, **19**, 1043 (2007).

50. A. V. Dmitriev, and I. P. Zvyagin, Current trends in the physics of thermoelectric materials. *Phys-Usp+*, **53**, 789 (2010).

51. D. L. Medlin, and G. J. Snyder, Interfaces in bulk thermoelectric materials are view for current opinion in colloid and interface science. *Curr. Opin. Colloid.* **14**, 226 (2009).

52. R. Y. Wang, R. A. Segalman, and A. Majumdar, Room temperature thermalconductance of alkanedithiol self-assembled monolayers. *Appl. Phys. Lett.*, **89**, 173113 (2006).

53. C. Yu, K. Choi, L. Yin, and J. C. Grunlan, Light-weight flexible carbon nanotube based organic composites with large thermoelectric power factors. *Acs. Nano.*, **5**, 7885 (2011).

54. W. L. Ong, S. M. Rupich, D. V. Talapin, A. J. H. McGaughey, and J. A. Malen, Surface chemistry mediates thermal transport in three-dimensional nanocrystal arrays. *Nat. Mater.*, **12**, 410 (2013).

55. S. Hida, T. Hori, T. Shiga, J. Elliott, and J. Shiomi, Thermal resistance and phonon scattering at the interface between carbon nanotube and amorphous polyethylene. *Int. J. Heat. MassTran.*, **67**, 1024 (2013).

56. S. K. Yee, N. E. Coates, A. Majumdar, J. J. Urban, and R. A. Segalman, Thermoelectric power factor optimization in PEDOT:PSS tellurium nanowire hybrid composites. *Phys. Chem. Chem. Phys.*, **15**, 4024 (2013).

57. N. G. Semaltianos, S. Logothetidis, N. Hastas, W. Perrie, S. Romani, R. J. Potter, G. Dearden, K. G. Watkins, P. French, and M. Sharp, Modification of the electrical properties of PEDOT:PSS by the incorporation of ZnO nanoparticles synthesized by laserablation. *Chem. Phys. Lett.*, **484**, 283 (2010).

58. B. Zhang, J. Sun, H. E. Katz, F. Fang, and R. L. Opila, Promising thermoelectric properties of commercial PEDOT:PSS materials and their Bi_2Te_3 powder composites. *ACS Appl. Mater. Inter.*, **2**, 3170 (2010).

Chapter 5

Role of Dopants in Defining Carrier Densities, Energetics, and Transport in Semiconducting Polymers

Gun-Ho Kim and Kevin P. Pipe

Dopants affect fundamental thermoelectric transport quantities including carrier concentration (n), carrier mobility (μ), and thermal conductivity (κ). While optimizing n by chemical doping is a well-known method to maximize the thermoelectric power factor ($S^2\sigma$, where S and σ are the Seebeck coefficient and electrical conductivity, respectively), the effect of finite dopant volume can greatly impact μ and hence $S^2\sigma$ in organic semiconductors (OSCs), since the hopping probability exponentially decreases with hopping distance. In this chapter, the role of dopants in tuning OSC thermoelectric transport parameters is discussed.

5.1 Introduction

Organic semiconductors are based on earth-abundant elements (e.g., C, H, O) and have many advantages over inorganic semiconductors (ISCs) such as low cost, low weight, and mechanical toughness. Their low thermal conductivity is beneficial for thermoelectric energy conversion efficiency, which is inversely proportional to κ (i.e., $\mathrm{ZT} = S^2\sigma T/\kappa$). In spite of these advantages, organic semiconductors (OSCs) have not traditionally been considered candidate thermoelectric materials, since they have historically been poor electrical conductors due to low charge carrier mobility. While ZT in OSCs had been typically three orders of magnitude smaller than that of heavy-metal alloys, it has rapidly improved since 2011, and now is reaching a

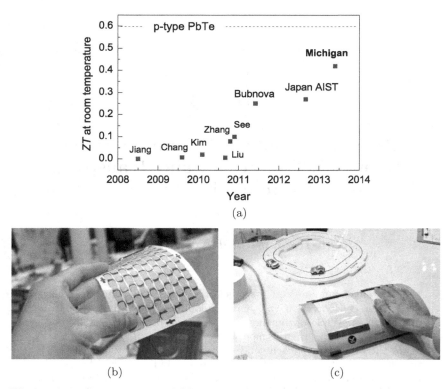

Figure 5.1. Current status of OSC-based thermoelectric materials. (a) Extraordinary recent progress in bulk (i.e., nontransistor) OSC-based TE materials (with respect to the common inorganic TE material PbTe). Data points appear in Refs. [2, 5, 6, 10, 32–35]. (b,c) Flexible OSC-based TE module developed by Fujifilm that runs a toy car using the heat from a human hand.[9] Japan AIST data and pictures are from Ref. [9].

value similar to that of the heavy-metal alloys, as shown in Fig. 5.1. Furthermore, the material properties governing thermoelectric efficiency (S, σ, and κ) do not have the same tradeoffs in OSCs that they have in crystalline ISCs, due to different transport mechanisms (e.g., hopping carrier transport), causing established property relationships such as the Wiedemann–Franz law to be invalid in OSCs.[1]

Poly(3,4-ethylenedioxythiophene) (PEDOT) is currently one of the most promising organic thermoelectric materials. ZT was derived to be 0.25 in chemically doped PEDOT thin films[2] and unity in electrochemically doped PEDOT transistors.[3] These high values are primarily due to PEDOT's large hole mobility, which is estimated to be $20 \, \text{cm}^2 \text{V}^{-1} \text{s}^{-1}$.[4] Mixtures of PEDOT with materials having large S (e.g., Te nanowires[5] or Bi_2Te_3 powders[6]) or large σ (e.g., carbon nanotubes[7] or graphene[8]) have demonstrated promising values of the thermoelectric power factor while κ was observed not to significantly increase.

High thermoelectric efficiency has recently been proven in an OSC at a device level (Fujifilm corp., March 2013)[9] through a demonstration in which a toy car ($\sim$10 mW) was operated by means of heat flux between a human hand and a heat sink at room temperature. Other recent work, discussed in this chapter, has demonstrated the importance of minimizing total dopant volume when maximizing the thermoelectric efficiency of OSCs.[10]

5.2 Optimal Carrier Concentration and Dopant Volume

In thermoelectric materials, the opposite dependences of S and σ on n lead to an optimal carrier concentration (n_{optimal}) that maximizes $S^2\sigma$. To realize n_{optimal}, the carrier concentration is tuned by means of chemical doping, which can affect μ in different ways depending on the nature of carrier transport. For crystalline ISCs, the carrier wave function is fully extended, and μ decreases with n because of increased impurity scattering at high dopant concentrations; this impurity scattering does not alter S significantly unless dopants resonate with the host semiconductor.[11] For OSCs in which the carrier wave function is localized, the dependence of μ on n is opposite; μ continuously increases with n until the detrimental effects of dopant volume on hopping distance[12] or the contributions of charge–dipole interactions to energetic disorder[13] become significant.

Figure 5.2 shows this strong dependence of μ on n in OSCs. Since the hopping probability exponentially decreases with hopping distance, a carrier localized by dynamic disorder at room temperature[14] tends to hop to the nearest vacant site. Because the average distance between states of a given energy is inversely related to the

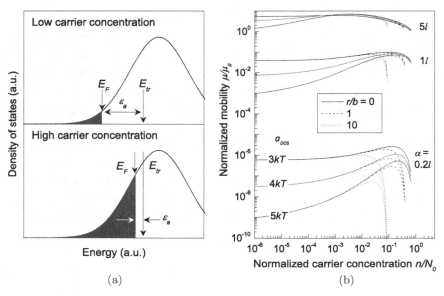

Figure 5.2. Hopping activation energy (ε_a), carrier mobility (μ), and carrier concentration (n). (a) Dependence of ε_a on n. (b) Calculated[12] mobility versus carrier concentration for various degrees of carrier localization (α), Gaussian DOS width (a_{DOS}), and nonionized dopant volume (r/b). While μ increases with n in general, a small b (or large r) leads to a profound decrease in μ at high n. This effect of r/b is greater for strongly localized carriers (i.e., small α).

cube root of the density of states (DOS) at that energy, the majority of nearby vacant sites are likely to be at energies nearer the center of the Gaussian DOS than its tail. As illustrated in Fig. 5.2(a), the difference between the average energy of the initial states ($\sim E_F$) and the final states ($\sim E_{tr}$) defines the hopping activation energy. As carrier concentration increases, E_F approaches E_{tr}, and the hopping activation energy becomes smaller, leading to a larger carrier mobility. Figure 5.2(b) shows the super-linear increase in the carrier mobility with n, the magnitude of which depends on the degree of carrier localization (α).

This strong mobility dependence in OSCs makes the dependence of $S^2\sigma$ on n very strong, and results in a much larger n_{optimal} for OSCs than ISCs (relative to the total DOS), as shown in Fig. 5.3. However, the inefficiency of dopant ionization in OSCs presents

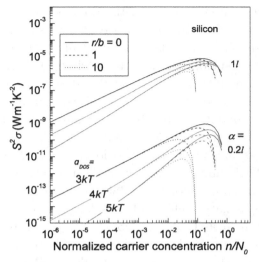

Figure 5.3. Thermoelectric power factor for various α and r/b. The strong dependence of $S^2\sigma$ on n and very large optimal carrier concentrations in OSCs are shown and compared with a typical ISC (silicon). Also shown is the significant effect of dopant volume (r/b) on $S^2\sigma$ in OSCs. $v_0 = 10^{13}\,\mathrm{s}^{-1}$ was used for OSCs.[12]

significant challenges to achieving these high concentrations without large negative effects on μ.

Inefficient dopant ionization arises in OSCs primarily due to a small dielectric constant relative to that of ISCs, causing charge carriers associated with dopants to be more bound to the dopantspecies. While the dopant activation energy in ISCs is typically less than kT at room temperature, it can be as large as several hundred meV in OSCs. At low carrier concentrations, it is on the order of the exciton binding energy. For a given dopant concentration n_d, the fraction of free charge carriers (i.e., those that contribute to electric conduction) is given by[15]

$$\frac{n}{n_d} = b = \exp\left(-\frac{E_a}{kT}\right) = \exp\left[-\frac{\left(E_{a,\max} - \beta n_d^{\frac{1}{3}}\right)}{kT}\right], \qquad (5.1)$$

where $E_{a,\max}$ is the maximum activation energy of the dopant at its dilute concentrations (approximately exciton binding energy) and β

captures the dependence of E_a on the dopant concentration. Given the fact that the exciton binding energy in OSCs ranges from 0.2 eV to 1.4 eV,[16] Eq. (5.1) predicts $b \ll 10^{-3}$ at dilute concentrations, indicating that most dopants do not contribute free carriers.

Due to this inefficiency of molecular doping, achieving high n often necessitates a composite material containing more electrically insulating dopants (>50 volume%) than electrically conductive host molecules.[17] For example, heavily doped pentacene can contain as many as 3.5 times as many iodine dopant atoms as pentacene molecules,[18] and a typical weight ratio between PEDOT and its dopantpoly(styrenesulfonate) (PSS) is 1:2.5, 1:6, or even 1:20[4] (Fig. 5.4). The negative effect of this large dopant fraction on μ and hence $S^2\sigma$ in heavily doped OSCs is expected to be especially significant for dopants with large size (e.g., $FeCl_4$ and molybdenum or cobalt organometallic complexes), while smaller atomic dopants such as F^-, Cl^-, and I^- may be too reactive to be practical.

Figure 5.2(b) illustrates the large effect of dopant volume on carrier mobility. While the carrier mobility tends to increase with carrier concentration, the inefficiency of molecular doping (i.e., $b \ll 1$) increases the dopant volume required to obtain a certain n and consequently creates a large reduction in carrier mobility at high dopant densities due to the presence of a high volume of insulating species.

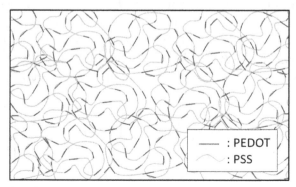

Figure 5.4. The ratio of PSS to PEDOT in typical PEDOT:PSS. A typical PEDOT:PSS mass ratio of 1:2.5 is used for this illustration.

5.3 Optimized Doping (or De-doping) of OSCs

Since charge carriers are spatially localized in OSCs due to either static (e.g., amorphous quality) or dynamic (e.g., thermal vibrations[19]) disorder and charge transport therefore occurs via thermally activated hopping,[20,21] the increase in distance between conducting molecules due to the presence of dopant molecules significantly decreases μ and consequently $S^2\sigma$. To quantify this effect due to finite dopant volume, the average distance that carriers at energy E must hop (tunnel) can be calculated[12]:

$$R(E) = \left(\frac{4\pi}{3B} \int_{-\infty}^{E} g(\varepsilon)(1 - f(\varepsilon))d\varepsilon \right)^{-\frac{1}{3}}, \qquad (5.2)$$

where B is the percolation threshold (2.7 for the case of 3D spheres[22]), $f(\varepsilon)$ is the Fermi–Dirac distribution, and $g(\varepsilon)$ is the Gaussian DOS:

$$g(\varepsilon) = \frac{N_t(r\chi)}{\sqrt{2\pi a_{\text{DOS}}^2}} \exp\left(-\frac{\varepsilon^2}{2a_{\text{DOS}}^2} \right) \qquad (5.3)$$

for which the total DOS (which depends on dopant volume) is:

$$N_t(r\chi) = \frac{N_0}{1 + r\chi} = \eta(r\chi)N_0 \qquad (5.4)$$

and N_0 is the total DOS of the undoped host material. The variable $r\chi$ is the ratio of total dopant volume to total host molecule volume, and is given by the product of the subunit (e.g., monomer for the polymer case) volume ratio (r) and the subunit number ratio (χ).

While the effect of dopant volume on N_t is negligible in ISCs, the relatively large dopant volume and inefficiency of molecular doping in OSCs lead to a large χ that significantly decreases N_t and hence increases the hopping distance (Eq. (5.2)). Using Eqs. (5.2)–(5.4), the dependence of $S^2\sigma$ on $r\chi$ and n can be calculated[12] and is illustrated in Fig. 5.5. The fact that the color contour is nearly circular means that the dependence of $S^2\sigma$ on $r\chi$ is just as strong in an OSC as it is

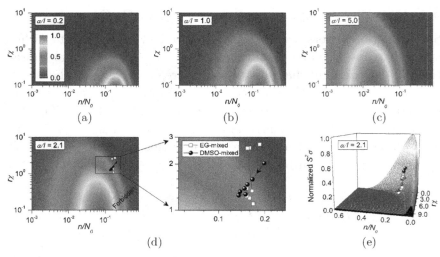

Figure 5.5. The 2D trajectory of doping (or de-doping) for efficient maximization of thermoelectric power factor in PEDOT:PSS. (a–c) Calculated[12] dependence of $S^2\sigma$ on the ratio of total dopant volume to total host volume ($r\chi$) and the normalized carrier concentration (n/N_0), for different degrees of carrier localization (α/l). Normalized $S^2\sigma$ is illustrated by the color bar. (d) Measured de-doping trajectory for DMSO-mixed (circles) and EG-mixed (squares) PEDOT:PSS. χ is determined by XPS data and n/N_0 is derived from numerical calculation[13] by fitting to measured data for S. The nonphysical condition of $n/N_0 > \chi$ for $r = 1.3$ is indicated by the forbidden area. The arrows in the magnified image represent the time evolution of the data during the EG treatment de-doping process. (e) Three-dimensional plot of normalized $S^2\sigma$, which demonstrates the steep ascent of the de-doping trajectory along both n/N_0 and $r\chi$ axes in the direction of maximum $S^2\sigma$. G.-H. Kim, L. Shao, K. Zhang, and K. P. Pipe, Engineered doping of organic semiconductors for enhanced thermoelectric efficiency. *Nat. Mater.*, **12**(8), 719–723 (2013).

on n/N_0, making $r\chi$ a primary engineering parameter for maximizing $S^2\sigma$. Furthermore, this plot shows the most efficient two-dimensional doping (or de-doping) trajectory to follow for the best possible $S^2\sigma$ in a particular OSC. For example, a doping strategy should focus on increasing b for an OSC located at the upper part of the color map (i.e., a vertical doping trajectory), whereas the carrier concentration should be optimized for an OSC located at the right or left part of the color map (i.e., a horizontal doping trajectory). For a heavily doped OSC located at the upper right part of the color map (such as PEDOT:PSS), the de-doping trajectory should be diagonal, not only optimizing n/N_0 but also largely reducing $r\chi$.

5.4 Effects of Reducing Dopant Volume on Thermoelectric Parameters

PEDOT is one of the most studied conducting polymers and is used in many applications such as transparent electrodes[23] and solar cell buffer layers[24] owing to its high electrical conductivity and stability in air. Most meaningful recent progress in organic thermoelectric materials depicted in Fig. 5.1 has arisen from PEDOT; $S^2\sigma$ equivalent to a ZT of 1 was measured in electrochemically modulated PEDOT transistors.[a,25]

PEDOT doped by PSS is available commercially in aqueous dispersions that yield films with large electrical conductivity ($>3000\,\mathrm{Scm^{-1}}$),[23] making it promising as a thermoelectric material. Pristine PEDOT:PSS is heavily doped, with ratios that range from 1:2.5 to 1:20. This abundant PSS increases the hopping distance between conducting PEDOT, and therefore exponentially reduces μ, making $S^2\sigma$ in pristine PEDOT:PSS very small.[26]

As Fig. 5.5(d) suggests, improving $S^2\sigma$ in PEDOT:PSS (in this case, Clevios PH1000 from H. C. Starck with a 1:2.5 mass ratio) requires reduction of both n and $r\chi$, which can be accomplished by careful de-doping of PSS. The hydrophilic nature of PSS allows a hydrophilic solvent such as ethylene glycol (EG) to separate PSS from PEDOT (which is hydrophobic).[24] We note that this EG treatment is separate from the mixing of EG or DMSO with PEDOT:PSS in solution; such chemical additives have been shown to significantly improve mobility in the resulting films. Figure 5.6a shows the decrease in PEDOT:PSS film thickness as an EG treatment procedure is applied. Selective removal of PSS was confirmed by separate XPS measurements, as the intensity of the sulfur atom (S_{2p}) from the sulfonate group in PSS (166–170 eV) decreased with respect to the intensity of the S_{2p} from the thiophene group (162–166 eV) in

---

[a]Bubnova *et al.* recently reported evidence of semimetallic behavior in PEDOT that supports high thermoelectric performance.

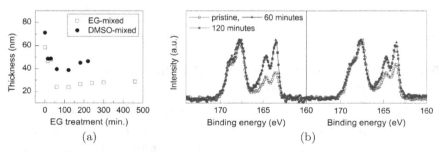

Figure 5.6. Selective removal of PSS by EG treatment. (a) Decrease in thickness in PEDOT:PSS with the EG treatment time. (b) S_{2p} XPS spectra in EG-mixed PEDOT:PSS (left) and DMSO-mixed PEDOT:PSS (right) at different EG treatment times. G.-H. Kim, L. Shao, K. Zhang, and K. P. Pipe, Engineered doping of organic semiconductors for enhanced thermoelectric efficiency. *Nat. Mater.*, **12**(8), 719–723 (2013).

PEDOT with longer EG treatment time (Fig. 5.6(b)). After 60 min of EG treatment, the relative fraction of PEDOT to PSS as measured by XPS became 1-to-1 and remained at this composition for longer EG treatment times, indicating that EG treatment no longer removed PSS. Presumably at this concentration the hydrophilic PSS chains become largely surrounded by hydrophobic PEDOT, making the hydrophilic EG solvent less effective in selectively removing PSS. Measurements of film thickness (Fig. 5.6(a)), XPS spectra (Fig. 5.6(b)), and Seebeck coefficient (Fig. 5.7(a)) each remained nearly constant for treatment times >60 min.

Approximating the volume ratio of the PSS monomer to the PEDOT monomer to be equal to their molecular weight ratio (PSS: 182, PEDOT: 140), $r\chi$ can be calculated at various points along the de-doping trajectory by means of XPS. The normalized carrier concentration (n/N_0) was obtained by fitting the measured S to numerical calculations.[12] The carrier localization length normalized by the molecular spacing (α/l) was set to 2.1, the value derived from other PEDOT:PSS data,[12,27] because its magnitude is not significantly affected by the dopant type or chemical additives.[27] As can be seen in Fig. 5.5(d), the de-doping trajectories for both EG-mixed and DMSO-mixed PEDOT:PSS aim directly at the maximum $S^2\sigma$, resulting in significant enhancements in $S^2\sigma$ (Figs. 5.5(e)

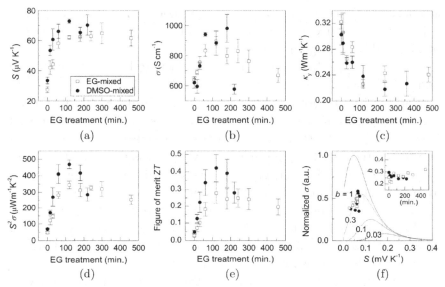

Figure 5.7. Thermoelectric properties of PEDOT:PSS at various de-doping times. (a) Seebeck coefficients, (b) electrical conductivities, (c) cross-plane thermal conductivities, (d) thermoelectric power factors, and (e) thermoelectric figure-of-merit at $T = 297$ K in EG-mixed and DMSO-mixed PEDOT:PSS. ZT was derived from in-plane thermal conductivities (same direction as measured $S^2\sigma$). The standard deviations of measured data from the linear fit line were used to determine error bars for S and σ. The standard deviations of measured temperature rise at different frequencies and thicknesses were used to determine error bars for κ_y. (f) Calculated[12] σ (normalized by a maximum value of $b = 1$) as a function of S for different magnitudes of b (assuming $r = 1.3$). Also shown are S and σ measured in EG-mixed and DMSO-mixed PEDOT:PSS. Inset: derived b during EG treatment. G.-H. Kim, L. Shao, K. Zhang, and K. P. Pipe, Engineered doping of organic semiconductors for enhanced thermoelectric efficiency. *Nat. Mater.*, **12**(8), 719–723 (2013).

and 5.7(d)). As the EG treatment was not able to reduce PSS below the 1-to-1 ratio, the de-doping trajectory stopped at $\chi = 0.96$ and $n/N_0 = 0.13$ for DMSO-mixed PEDOT:PSS; at this point, $S^2\sigma$ reaches $469\,\mu\mathrm{Wm^{-1}K^{-2}}$, which is approximately half of the calculated maximum $S^2\sigma$. If n/N_0 can be reduced to 0.07 and $b \sim 1$ (i.e., $r\chi$ is minimized), $S^2\sigma$ is predicted to be as high as $1{,}100\,\mu\mathrm{Wm^{-1}K^{-2}}$, which is close to the measured $S^2\sigma = 1{,}270\,\mu\mathrm{Wm^{-1}K^{-2}}$ in electrochemically modulated PEDOT transistors.[3] Although transistor geometries are not practical as thermoelectric devices, achieving such a high performance in a bulk OSC (which would bring ZT close to 1) could have large impact.

In OSCs, $n_{optimal}$ is very large ($\sim 0.1 N_0$), and the typically large number of dopants present due to small b (Eq. (5.1)) strongly affects μ (Fig. 5.2(b)). In both EG-mixed and DMSO-mixed PEDOT:PSS, S and σ were found to simultaneously increase as PSS was removed (Figs. 5.7(a) and 5.7(b)), indicating that the significant increase in μ due to the reduced hopping distance overwhelms the reduction in n. While $d\sigma/dS$ is typically negative in ISCs,[28] this unique trend of positive $d\sigma/dS$ occurs at high doping concentrations for OSCs with large $n_{optimal}$, small b, and large molecular dopant volume (Fig. 5.7(f)).

In addition to this simultaneous increase in S and σ with de-doping, the cross-plane thermal conductivity κ_y was found to decrease with decreasing PSS concentration (Fig. 5.7c). Measured κ_y in the pristine EG-mixed and DMSO-mixed PEDOT:PSS were $0.32\,\mathrm{Wm^{-1}K^{-1}}$ and $0.30\,\mathrm{Wm^{-1}K^{-1}}$, respectively, which decreased to $0.23\,\mathrm{Wm^{-1}K^{-1}}$ and $0.22\,\mathrm{Wm^{-1}K^{-1}}$, respectively, as PSS de-doping progressed. This decrease in κ is a scribed to the much larger molecular weight of PSS ($\sim 400{,}000\,M_w$) versus PEDOT ($\sim 2{,}000\,M_w$) and to changes in nanoscale morphology. The much larger molecular weight of PSS implies that PEDOT:PSS with less PSS contains a greater number ratio of VDW bonds to covalent bonds. Since the bonding strength of the VDW bond is much weaker than the covalent bond, κ in PEDOT:PSS with less PSS is expected to be smaller. Additionally, PEDOT:PSS films are known to form nanostructures in which PEDOT-rich regions are surrounded by PSS-rich shells.[29] During PSS removal, the thicknesses of the PSS-rich shells likely decrease, leading to a larger nano-interface density and hence smaller κ.

The derived anisotropy ratio κ_x/κ_y was used to convert κ_y measured at different doping levels to κ_x, which was then used to derive ZT. Using the three thermoelectric parameters that were measured in the in-plane direction, a maximum ZT of 0.42 was derived for the DMSO-mixed PEDOT:PSS and 0.28 for the EG-mixed PEDOT:PSS, the former being the highest ZT yet reported among OSCs. This high value of ZT suggests the importance of minimizing dopant volume

$(r\chi)$ in maximizing ZT, which has been shown above to simultaneously enhance all three parameters constituting ZT.

Other p-type organic thermoelectric materials that can be doped to high concentrations are expected to have similar effects related to dopant volume. For example, a substantial number of tosylate (Tos) dopants in PEDOT[2] do not contribute free charge carriers at the optimized oxidation level (e.g., $b \sim 0.1$[12]), suggesting that the realized maximum ZT = 0.25 in PEDOT:Tos can be further improved by removing excess dopants. Likewise, $S^2\sigma$ in poly(2,7-carbazole) doped by FeCl$_4$ was observed to be optimized near a 1-to-1 ratio between the monomer unit and FeCl$_4$,[30] suggesting that a large number of excess dopants exist, which may be removed to further improve $S^2\sigma$. On the other hand, n-type doping is rare in OSCs, since OSCs typically have a large bandgap which places their LUMO (lowest unoccupied molecular orbital) at a high energy. Efficient n-type dopants must have a HOMO (highest occupied molecular orbital) level higher than the LUMO of the host OSC, making n-type dopants difficult to achieve, which are stable in ambient conditions. Currently, OSCs must have low band gap (e.g., fullerene) in order to be n-type doped. Only one report has yet shown a meaningful ZT in an n-type OSC.[31]

5.5 Conclusions

Minimizing dopant volume can induce unique trends in OSC thermoelectric parameters (S and σ increase while κ decreases) and therefore is a primary engineering guideline when maximizing ZT. Recent work in optimizing an OSC's oxidation level[2] and developing polymer-CNT composites[7] and flexible organic thermoelectric devices[9] suggest a bright future for organic thermoelectric materials. Since organic materials are lightweight, inexpensive, mechanically tough and capable of large area deposition, they are strong candidates for thermoelectric energy conversion near room temperature.

Acknowledgments

This work was supported as part of the Center for Solar and Thermal Energy Conversion, an Energy Frontier Research Center funded by the US Department of Energy, Office of Science, Basic Energy Sciences under Award No. DE-SC0000957.

References

1. A. Casian, Violation of the Wiedemann–Franz law in quasi-one-dimensional organic crystals. *Phys. Rev. B*, **81**(15), 155415 (2010).
2. O. Bubnova, Z. U. Khan, A. Malti, *et al.*, Optimization of the thermoelectric figure of merit in the conducting polymer poly(3,4-ethylenedioxythiophene). *Nat. Mater.*, **10**(6), 429–433 (2011).
3. T. Park, C. Park, B. Kim, *et al.*, Flexible PEDOT electrodes with large thermoelectric power factors to generate electricity by the touch of fingertips. *Energy Environ. Sci.*, **6**(3), 788–792 (2013).
4. A. Elschner, *PEDOT: Principles and Applications of an Intrinsically Conductive Polymer*. CRC Press: Boca Raton, FL, xxi, p. 355, 2011.
5. K. C. See, J. P. Feser, C. E. Chen, *et al.*, Water-processable polymer-nanocrystal hybrids for thermoelectrics. *Nano Lett.*, **10**(11), 4664–4667 (2010).
6. B. Zhang, J. Sun, H. E. Katx, *et al.*, Promising thermoelectric properties of commercial PEDOT:PSS materials and their Bi2Te3 powder composites. *Acs Appl. Mater. Interfac.*, **2**(11), 3170–3178 (2010).
7. P. Gregory, K. B. Moriarty, B. Stevens, C. Yu, and J. C. Grunlan, Fully organic nanocomposites with high thermoelectric power factors by using a dual-stabilizer preparation. *Energy Technol.*, **1**(4), 265–272 (2013).
8. K. Zhang, Y. Zhang, and S. R. Wang, Enhancing thermoelectric properties of organic composites through hierarchical nanostructures. *Sci. Rep.*, **3**, 3448 (2013).
9. H. Kume, Fujifilm shows high-efficiency thermoelectric converter using organic material. 2013 [cited 2013 October 19]; Available from: http://techon.nikkeibp.co.jp/english/NEWS_EN/20130206/264517.
10. G.-H. Kim, L. Shao, K. Zhang, and K. P. Pipe, Engineered doping of organic semiconductors for enhanced thermoelectric efficiency. *Nat. Mater.*, **12**(8), 719–723 (2013).
11. J. P. Heremans, V. Jovovic, E. S. Toberer, *et al.*, Enhancement of thermoelectric efficiency in PbTe by distortion of the electronic density of states. *Science*, **321**(5888), 554–557 (2008).
12. G. Kim and K. P. Pipe, Thermoelectric model to characterize carrier transport in organic semiconductors. *Phys. Rev. B*, **86**(8), 085208 (2012).
13. D. H. Dunlap, P. E. Parris, and V. M. Kenkre, Charge-dipole model for the universal field dependence of mobilities in molecularly doped polymers. *Phys. Rev. Lett.*, **77**(3), 542–545 (1996).

14. J. Bohlin, M. Linares, and S. Stafstrom, Effect of dynamic disorder on charge transport along a pentacene chain. *Phys. Rev. B*, **83**(8), 085209 (2011).

15. B. A. Gregg, S. G. Chen, and R. A. Cormier, Coulomb forces and doping in organic semiconductors. *Chem. Mater.*, **16**(23), 4586–4599 (2004).

16. M. Knupfer, Exciton binding energies in organic semiconductors. *Appl. Phys. A — Mater. Sci. Process.*, **77**(5), 623–626 (2003).

17. K. Walzer, B. Maennig, M. Pfeiffer, *et al.*, Highly efficient organic devices based on electrically doped transport layers. *Chem. Rev.*, **107**(4), 1233–1271 (2007).

18. M. Brinkmann, V. S. Videva, A. Bieber, *et al.*, Electronic and structural evidences for charge transfer and localization in iodine-doped pentacene. *J. Phys. Chem. A*, **108**(40), 8170–8179 (2004).

19. A. Troisi and G. Orlandi, Dynamics of the intermolecular transfer integral in crystalline organic semiconductors. *J. Phys. Chem. A*, **110**(11), 4065–4070 (2006).

20. R. Coehoorn, W. F. Pasveer, P. A. Bobbert, *et al.*, Charge-carrier concentration dependence of the hopping mobility in organic materials with Gaussian disorder. *Phys. Rev. B*, **72**(15), 155206 (2005).

21. N. F. Mott and E. A. Davis, Electronic processes in non-crystalline materials. *International Series of Monographs on Physics*. Clarendon Press: Oxford, xiii, p. 437, 1971.

22. D. Stauffer and A. Aharony, *Introduction to Percolation Theory*. 2nd ed. Taylor & Francis: London, Washington, DC, x, p. 181, 1992.

23. Y. J. Xia, K. Sun, and J. Y. Ouyang, Solution-processed metallic conducting polymer films as transparent electrode of opto electronic devices. *Adv. Mater.*, **24**(18), 2436–2440 (2012).

24. Y. H. Kim, C. Sachse, M. L. Machala, *et al.*, Highly conductive PEDOT:PSS electrode with optimized solvent and thermal post-treatment for ITO-free organic solar cells. *Adv. Funct. Mater.*, **21**(6), 1076–1081 (2011).

25. O. Bubnova, Z. U. Khan, H. Wang, *et al.*, Semi-metallic polymers. *Nat. Mater.* (Advanced Online Publication) (2013).

26. R. R. Yue and J. K. Xu, Poly(3,4-ethylenedioxythiophene) as promising organic thermoelectric materials: A mini-review. *Synthetic Metals*, **162**(11–12), 912–917 (2012).

27. P. K. Kahol, K. K. S. Kumar, S. Geetha, *et al.*, Effect of dopants on electron localization length in polyaniline. *Synthetic Metals*, **139**(2), 191–200 (2003).

28. G. J. Snyder and E. S. Toberer, Complex thermoelectric materials. *Nat. Mater.*, **7**(2), 105–114 (2008).

29. U. Lang, E. Muller, N. Haujoks, *et al.*, Microscopical investigations of PEDOT:PSS thin films. *Adv. Funct. Mater.*, **19**(8), 1215–1220 (2009).

30. R. B. Aich, N. Blouin, A. Bouchard, *et al.*, Electrical and thermoelectric properties of poly(2,7-carbazole) derivatives. *Chem. Mater.*, **21**(4), 751–757 (2009).

31. Y. M. Sun, P. Sheng, C. A. Di, *et al.*, Organic thermoelectric materials and devices based on p- and n-type poly(metal 1,1,2,2-ethenetetrathiolate)s. *Adv. Mater.*, **24**(7), 932 (2012).

32. F. X. Jiang, K. K. Xu, B. Yu, *et al.*, Thermoelectric performance of poly(3,4-ethylenedioxythiophene): Poly(styrenesulfonate). *Chin. Phys. Lett.*, **25**(6), 2202–2205 (2008).

33. K. C. Chang, M. S. Jeng, C. C. Yang, *et al.*, The thermoelectric performance of poly(3,4-ethylenedioxythiophene)/poly(4-styrenesulfonate) thin films. *J. Electron. Mater.*, **38**(7), 1182–1188 (2009).

34. D. Kim, Y. Kim, K. Choi, *et al.*, Improved thermoelectric behavior of nanotube-filled polymer composites with poly(3,4-ethylenedioxythiophene) poly(styrenesulfonate). *Acs Nano*, **4**(1), 513–523 (2010).

35. C. C. Liu, B. Y. Liu, J. Yan, *et al.*, Highly conducting free-standing poly(3,4-ethylenedioxythiophene)/poly(styrenesulfonate) films with improved thermoelectric performances. *Synthetic Metals*, **160**(23–24), 2481–2485 (2010).

Chapter 6

Thermoelectric Polymer–Inorganic Composites

*Robert M. Ireland and Howard E. Katz**

Polymer–inorganic composites have become widely anticipated and explored for thermoelectric (TE) applications, particularly with the parallel development of conducting organic polymers and compound semiconductor nanostructures. Physical mixing, solution processing, and *in situ* synthesis of polymers and/or inorganic crystals have been used to prepare polymer–inorganic composites which show TE improvements over the individual constituents, leading to useful power output and high practical efficiencies. The incorporation of conducting inorganic compounds in conjugated polymers is expected to increase composite electrical conductivity and can also enhance the Seebeck coefficient, while the polymer matrix retains a very low thermal conductivity. Composites are expected to have improved TE performance due to enhanced current pathways, donation or extraction of carriers, morphological confinement, passivation or functionalization, energy alignment and filtering, and interface scattering effects.

6.1 Introduction

Although organic polymers are generally not stable at high temperatures (above 300°C), they could still be of interest in niche thermoelectric (TE) applications for room temperature cooling and power generation on the microscale where sufficient power is more critical than high efficiency. Their low processing cost and environmental impact compared to typical TE materials will allow sustainable

---

*Howard Katz is grateful to the Department of Energy, Office of Basic Energy Sciences, Grant Number DE-FG02-07ER46465, for supporting his contributions.

147

large-scale production. Applications include spot cooling, small-scale refrigeration, thermal diodes, and off-grid or battery-free power supply to wireless or mobile devices (i.e., autonomous sensors, medical implants).

Inorganic material fabrication as bulk solids involves high-cost processes, making it challenging to affordably integrate these rigid, high-performance materials into large-area modules or onto unusual topologies. Hence, there is growing and immediate interest in composites of inorganic particles and organic semiconducting materials for thermoelectric energy harvesting and spot cooling applications.[1-26] Such composites combine solution processing, mechanical flexibility, and potentially low thermal conductivity with high electrical conductivity, contributing to high values of the figure of merit, $ZT = S^2\sigma T/\kappa$, where S is the Seebeck coefficient, σ is the electronic conductivity, T is the absolute temperature, and κ is the thermal conductivity.[21,22] $S^2\sigma$ is known as the power factor PF. Clearly a good TE efficiency requires a high S coefficient, high σ, and low κ. This is beneficial for TE materials operating in steady state because a high-performance material is able to hold a large temperature difference in order to maintain the S voltage, which should respond strongly to small T gradients, but also needs good short-circuit current to be useful in devices. The most attractive feature of organic materials is their low κ ($<0.5\,\mathrm{W\,m^{-1}\,K^{-1}}$) compared to most inorganics ($>1\,\mathrm{W\,m^{-1}\,K^{-1}}$). Low κ in polymers is due to van der Waals interactions between molecules and long-range structural disorder, and the σ in polymers is too low for the electron contribution, κ_e, to become significant.[23-25] Therefore, one major challenge for organic materials is to maximize PF, which may be most easily accomplished through optimal loading with inorganic inclusions.[16,26] Figure 6.1 shows the side view schematic of measurement configurations used to analyze TE performance, including lateral (in-plane, Fig. 6.1(a)) and vertical (through-plane, Fig. 6.1(b)) thin films (50–5,000 nm thick). A temperature gradient is generated using heating elements (i.e., Peltier tiles) and that the temperature difference and open-circuit voltage, V_{OC}, are measured across the film between electrodes.

Polymer-based composites would be less expensive to implement, could be applied to surfaces with irregular geometries, and would

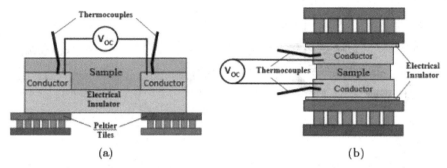

Figure 6.1. Illustrations depicting the side view of (a) lateral and (b) vertical configurations used to measure TE performance of thin films.

desirably contain only common and nontoxic elements. Polymers have inherent advantages in their physical and mechanical properties, such as light weight, tunable flexibility, and broad optical modification, which contribute practical efficiency to polymer-based electronics rather than performance efficiency.[27,28] Furthermore, the versatility of polymer processing, and low energy required to make these materials, renders them more cost-effective and allows alternative processing techniques to be utilized, such as all-solution processing and printing techniques, which have been shown to be the preferred methods for making cost-effective devices at scale.[29–31] Although focus remains on improving the thermoelectric efficiency of particular materials, there have been a handful of demonstrations showing thermoelectric generator (TEG) prototypes based on organic polymers. Figure 6.2 shows examples of previously reported device architectures. Devices exhibit lateral (Figs. 6.2(a) and 6.2(c))[8,35] and vertical (Figs. 6.2(b) and 6.2(d))[30,41] architectures, or both (Figs. 6.2(e) and 6.2(f)),[42,126] achieved by solution-casting techniques, vapor deposition, or by pressing insoluble solutions or powders into pellets.

Significant progress has been made toward increasing thermoelectric PF of hole-conducting (p-type) polymers,[1–24,26,29–32,43–49] with PF values $>100\,\mu\mathrm{W\,m^{-1}\,K^{-2}}$ that lead to $ZT > 0.1$. Figure 6.3 shows the chemical structures for polymers and small molecules discussed herein. Although polyacetylene (PA, Fig. 6.3(a)) achieves very high values of σ, resulting in reported ZT as high as 0.1–0.38,[49] it is

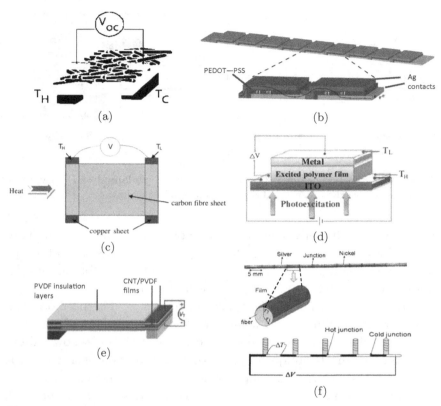

Figure 6.2. Examples of previously reported thermoelectric modules and microlegs made from polymer-based materials. Images (a): adapted with permission from See *et al.*[8] Copyright (2010) American Chemical Society. Image (b): adapted with permission from Sondergaard *et al.*[30] Copyright (2013) Wiley. Image (c): reprinted with permission from Pang *et al.*[35] Copyright (2011) Elsevier. Image (d): reprinted with permission from Xu *et al.*[41] Copyright (2013) American Chemical Society. Image (e): adapted with permission from Hewitt *et al.*[42] Copyright (2012) American Chemical Society. Image (f): adapted with permission from Yadav *et al.*[126] Copyright (2008) Elsevier.

unstable to oxidation in atmosphere. An alternative polymer with good conductivity and environmental stability is polyaniline (PANI, Fig. 6.3(e)), which has been the most studied TE polymer. PANI forms metallic domains with good σ between more disordered regions that contribute to low thermal conductivity.[50] Charge is transported between metallic domains by hopping among interconnecting

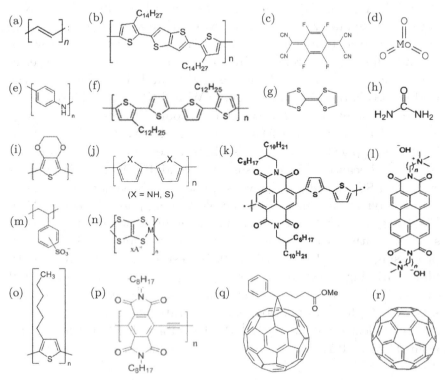

Figure 6.3. Chemical structures of (a) PA, (b) PBTTT-C14, (c) F4TCNQ, (d) MoO₃, (e) PANI, (f) PQT12, (g) TTF, (h) urea, (i) PEDOT, (j) PTH X = S and PPY X = NH, (k) P(NDI2OD-T2), (l) self-doped perylene diimide, (m) PSS, (n) organometallic poly(Ni 1,1,2,2-ethenetetrathiolate), (o) P3HT, (p) PyDI-5FPE, (q) PCBM, and (r) C₆₀.

conductive chains. Thusly, the TE properties are strongly dependent on preparation conditions and methods, and on molecular weight, oxidation level, doping level, molecular morphology, crystallinity, and inter-chain distances.[51] The TE performance of PANI materials has been enhanced most in the case of nanotubes,[52] multilayed films,[53,54] and stretched films.[55] PANI exhibits thermal conductivity from 0.02 W m⁻¹ K⁻¹ to 0.54 W m⁻¹ K⁻¹, and has achieved a ZT of 0.011 at 423 K.[52–54]

The highest TE performance obtained has been from the conducting thiophene polymer poly(3,4-ethylenedioxythiophene)

(PEDOT, Fig. 6.3(i)), which has reached PF of $469\,\mu W\,m^{-1}\,K^{-2}$ with an ultimate ZT of 0.42 at room temperature upon doping with poly(styrenesulfonic) acid (PSS, Fig. 6.3(j)).[24] In devices where PEDOT films were prepared by solution casting with precise electrochemical control of polymer oxidation and doping with tosylic acid (TOS), a power factor of $1,270\,\mu W\,m^{-1}\,K^{-2}$ was achieved at an electrode potential of 0.1 V, which corresponds to a ZT approaching a value of 1.[32] However, PEDOT has limited solvent solubility which limits its application. It can also have a considerably complicated structure because it is necessary to blend PEDOT with dopants such as PSS in order to have semimetallic behavior. Thiophene-based polymers in general, the simplest being polythiophene (PTH, Fig. 6.3(j) and $X = S$), are well studied and are now considered model systems, with a wide range of high σ upon doping and optimized processing $(0.001–1000\,S\,cm^{-1})$. It is desirable to prepare smooth films with high crystallinity, compact packing, and good orbital overlap.[56–62] Gao *et al.* predicted high p- and n-type S (100 and $-140\,V\,K^{-1})$ for very low doping concentrations (0.04 hole/monomer and 0.02 electron/monomer).[63] PTH exhibits thermal conductivity from 0.028 to $0.17\,W\,m^{-1}\,K^{-1}$, and has achieved a ZT of 0.029 at 250 K.[64]

Favorable electronic density of states should be designed in order to improve TE performance. Katz *et al.* showed that the rational design of the density of states could lead to a simultaneous increase in S and σ by introducing a small amount of more easily doped conductive polymer into the matrix of less easily doped polymer.[48] Specifically, the dopant concentration using tetrafluorotetracyanoquinodimethane (F4TCNQ, Fig. 6.3(c)) was increased from 0.2 to 0.7 wt% within a composite composed of 8 wt% poly(3-hexylthiothiophene) (P3HTT) in poly(3-hexylthiophene (P3HT, Fig. 6.3(m)). S and σ both increased from 460 to $530\,\mu V\,K^{-1}$ and 2×10^{-5} to $3.7 \times 10^{-5}\,S\,cm^{-1}$, respectively. S was higher for composites with lower P3HTTT (2 wt%) and F4TCNQ concentrations (0.25 wt%), reaching $700\,\mu V\,K^{-1}$. The PF increased from 4.62×10^{-4} to $7.58 \times 10^{-3}\,W\,m^{-1}\,K^{-2}$, indicating that organic-based compositions may be designed with intentional inhomogeneity based on the theoretically preferred density of states distributions for TE materials.

Unfortunately, it is more difficult to make n-type organic materials due to their generally lower electron affinities relative to the environment, and the search for high-efficiency n-type polymer-based materials is more challenging. N-type materials are required, however, in order to make a most efficient TEG. Of the currently available n-type materials, fullerenes (Figs. 6.3(o) and 6.3(p))[65–68] and powder-processed organometallic poly(Ni 1,1,2,2-ethenetetra-thiolate) (Fig. 6.3(g))[38] derivatives have shown high thermoelectric performance, with electrical conductivities reaching 9 and 40 S cm^{-1}, resulting in power factors of up to 30 μW m^{-1} and 70 μW m^{-1} K^{-2}, respectively. However, these materials are not amenable to solution processing. The few results obtained regarding solution-processable materials have been on imide-containing polymers,[69,70] or imide-based small molecules.[71] The first was an all solution-processable n-type polymer having very simple and rigid planar chemical structure (Fig. 6.3(n)). S was reported to be around $-40\,\mu$V K^{-1}.[69] Following that, Schlitz *et al.* demonstrated solution doping of a high-mobility n-type polymer, poly[N,N'-bis(2-octyl-dodecyl)-1,4,5,8-napthalenedicarboximide-2,6-diyl]-alt-5,5$'$-(2,2$'$-bithiophene)] (P(NDI2OD-T2), Fig. 6.3(k)), using dihydro-1H-benzoimidazol-2-yl derivatives as potential dopants.[70] The authors achieved electrical conductivities of nearly 0.01 S cm^{-1} and PF of 0.6 W m^{-1} K^{-2}. Segalman and co-workers showed record thermoelectric performance for solution-processed perylene diimide molecules, which could be designed with doping atoms separated by spacer carbon atoms, resulting in σ of 0.4 S cm^{-1}, S of around $-180\,\mu$V K^{-1}, and PF of 1.4 μW m^{-1} K^{-2} (Fig. 6.3(l)).[71]

In the case of metal coordination n- and p-type organometallic polymers,[38] Zhu *et al.* observe an increase in ZT with temperature for all such materials investigated, where conductivity versus temperature indicates charge transport is three-dimensional variable range hopping (3D-VRH), typical of high σ polymers, whereas lower conductivity films usually exhibit nearest-neighbor distance hopping (NDH). Similarly, carrier transport in thin films of inorganic nanocrystals occurs via hopping, a thermally activated process, which presents space-charge limitations due to accumulation of slow carriers.[186,187] The linear increase in S with temperature for coordination polymers is similar to doped conducting polymers such as PTH and polypyrrole (PPY, Fig. 6.3(j) and X = NH). A U-shaped temperature dependence in S indicates possible tunneling between granular metallic domains, like in PANI.[72] For the highest performing coordination polymer, poly[Nax(Ni-ett)], power factors range from $6 \mu W \, cm^{-1} \, K^{-2}$ to $60 \mu W \, cm^{-1} \, K^{-2}$ (S around -100–$-150 \mu V \, K^{-1}$, and conductivities of 5–$40 \, S \, cm^{-1}$), leading to ZT of 0.1–0.2 at 300–$400 \, K$, though the materials are completely insoluble. N-type poly[Nax(Ni-ett)] and p-type poly[Cux(Cu-ett)] were combined to make TEGs which exhibit a power of $750 \mu W$ utilizing 35 p–n couples. The polymers were pressed ($2 \, MPa$) into pellets to form the individual p- or n-legs typical of conventional TEGs. Interestingly, the conductivity of poly[Cux(Cu-ett)] is about 50 times higher than the other p-type material investigated, poly[Nax(Cu-ett)], while they show similar S coefficients of about $80 \mu V \, K^{-1}$. The σ of n-type polymers ($40 \, S \, cm^{-1}$ at $300 \, K$) was more than double that of poly[Cux(Cu-ett)] ($15 \, S \, cm^{-1}$ at $300 \, K$) over the temperature range of 220 to $440 \, K$, while the thermal conductivity of n-type ($0.2 \, W \, m^{-1} \, K^{-1}$ at $300 \, K$) was about half compared to that of p-type ($0.35 \, S \, cm^{-1}$ at $300 \, K$). One possible reason for unipolar transport in Ni- and Cu-based coordination polymers is that oxidation may selectively occur on either center metal atoms or ligand bridges depending on the species of the central metal atoms, similar to what is observed in metal phthalocyanine systems.

6.2 Morphological Considerations: Polymers

There is no obligation to follow the same strategies as pure inorganics regarding improved TE performance in conjugated polymers. For one, the average mean free path of phonons in polymers is expected to be much smaller than $10 \, nm$ at room temperature,[73] but too long for incipient localization to have such a large effect in homogeneous systems/models.[74] Polymers exhibit narrow to moderate band gaps between the lowest unoccupied molecular orbital (LUMO, conduction band with states trailing into the gap) and the highest occupied molecular orbital (HOMO, valence band with states trailing into the gap), around 1–3 eV. Conducting polymers also exhibit dissimilar carrier transport and relaxation kinetics compared to inorganic materials, and have intrinsically lower carrier motilities (typically much less than $1 \, cm^2 \, V^{-1} \, s^{-1}$, mostly due to morphological disorder), and much lower thermal conductivity compared to inorganic materials.

Significantly, many polymers are anisotropic in their conductivity on the microscale due to their 1D periodic chain-like structures. Modern conducting polymers can even have stable σ in air on par with the best inorganic TE materials (PA, PANI, PTH, PEDOT:PSS, and Bi_2Te_3 ~1,000 S cm^{-1}). Organic conducting polymers exhibit surprisingly large conductivities for materials with low carrier density and high disorder. Conducting polymers can thus be thought of as phonon-glass electron-crystal (PGEC) materials which have good σ like a crystalline conductor and low thermal conductivity like an amorphous glass.[76] However, macromolecular conjugated systems are very complex in morphology since their structure is typically associated with randomly coiled chains of atoms, and have distributions in density and free volume, forming crystalline domains separated by amorphous regions, which ultimately require overlapping conjugated cores, and thus, overlapping orbitals for electrical conduction.[77] Polymers generally form semicrystalline films depending on the structure and ability to pack or form large conjugated regions, resulting in some degree or percentage of crystallinity (can be as high as 70%–90% in P3HT), with the size of crystallites of ~10–20 nm in diameter.[78] Although conjugated polymers are very complicated structurally, this material system, particularly as a host material or interfacial film for other conductive TE inclusions, may allow for decoupling between TE parameters that are traditionally interdependent, particularly for inorganic materials, such as κ and σ. Heterogeneous structures are known to contribute decoupling of the TE parameters in hybrid materials.[16,36,39–41,52–55,74,79–85,89–95,98,110–114]

At first glance it may seem remarkable that the greater enhancements achieved in the TE properties of polymeric materials (including many polymer–inorganic composites) do not involve nanostructuring, which has improved inorganic films. This can be expected since the main accomplishment of reducing dimensionality in traditional materials has been the reduction of the thermal conductivity rather than enhancement of the power factor due to quantum confinement effects. Considering that organic compounds have intrinsically low values of κ, little benefit could be expected from attempting to reduce it. In general, the transport path of the charge carriers in polymeric materials involves more than one molecular chain, so that the conduction is not so much controlled by the properties within the molecular chain, but by charge transfer between chains.[77] Nanostructuring could make percolation more difficult to achieve in bulk films at low concentrations. The κ is typically decreased when there are more interfaces present, due to increased boundaries from nanostructuring or introducing impurities, which reduce the free path distance of phonons. Enhanced S and σ in the case of nanostructured polymeric materials typically arise from the realignment of the molecular chains, as in the case of properly doped films,[80–85] multilayerd films,[53,54] stretched polymer films,[55] structural anisotropy in the bulk,[89–91] polymer nanotubes/wires/crystals,[52,92,93] molecular microwires,[94] and n-type polymer fibers.[95]

X-ray diffraction patterns and UV–vis spectra of (±)-10-camphorsulfonic acid (CSA)-doped stretched PANI films suggested that an increase in σ can be attributed to the induced alignment of PANI chains into coil-like conformations, which increases the carrier mobility. It is also found that stretching can increase the σ parallel to the stretched direction to twice that of the perpendicular direction.[55] Similarly, the S coefficient parallel to the stretched direction is also higher than that perpendicular to that direction. In CSA-doped PANI, the bulk ZT value exhibited a sixfold increase (to 0.011 at 423 K) when prepared in a multilayered film structure composed of electrically insulating and electrically conducting layers.[53,54] Protonic acid-doped PANI films exhibit low thermal conductivity (~ 0.1–$0.2\ \mathrm{W\,m^{-1}\,K^{-1}}$) that is not correlated with either σ (greatest value $\sim 200\ \mathrm{S\,cm^{-1}}$) or structure of proton acid dopant, i.e., the thermal conductivity was not systematically investigated with regard to dopant species or as a function of doping concentration.[80] The comparison of such films with acid-doped PANI nanotubes shows that molecular ordering of polymer chains enhances charge carrier mobility in a way that both S and σ simultaneously increase,[52] similar to stretched films.

It was shown experimentally that PEDOT:PSS films mixedwith urea (Fig. 6.3(h)) have an altered structure compared to neat films, leading to a simultaneous enhancement in σ and S.[81] Urea does not increase carrier concentration but apparently increases carrier mobility. PEDOT:PSS films mixed with the polar solvent dimethyl sulfoxide (DMSO) show a change of structure from coiled to linearly expanded-coil conformation, for both oxidatively polymerized films and films prepared by spin-casting.[82–85] In this case, the σ between PEDOT components is enhanced while S is essentially unaffected, although no increase in carrier concentration is expected.

It is found that the molecular weights of polymers have a substantial effect on the electron mobility and consequently affect σ. Kline *et al.* found that an increase in the chain length (molecular weight) of PTH polymers leads to the increase in electron mobility through electron hopping along the polymer backbone.[86] A high molecular weight polymer will promote the charge carriers to move longer distances before hopping to another chain, and longer chains give carriers more opportunities for hopping to neighboring chains. Also, Zhao *et al.* show that the thermal conductivity of amorphous polyethylene increases with the increase of polymer chain length, which follows the conventional relationship between κ and σ.[87]

The nonfluorinated moiety of F4TCNQ can form a charge transfer salt with tetrathiafulvalene (TTF, (Fig. 6.3(g)), resulting in a complex of two planar organic molecules bonded by overlapping pi orbitals. The complex was the first organic compound observed to have σ that is anisotropic or quasi-one-dimensional due to the overlap of molecular orbitals along stacked chains compared to conductivity between chains, having the notorious "herring bone" crystal structure.[88–90] σ was theoretically shown to be significantly higher along certain packing directions, and the sign of the S coefficient can be negative or positive depending on the orientation of the measurement with respect to the crystal packing direction. P-type TTF-TCNQ thin films show an experimental PF of 0.78 μWm^{-1} K^{-2}.[91]

Penner *et al.* prepared PEDOT nanowires (40–90 nm height, 150–580 nm width, and about 200 μm length) by electropolymerizing EDOT in aqueous LiClO$_4$ using the lithographically patterned nanowire electrodeposition process.[92] Ultralong PEDOT nanowires exhibit S ($-122\,\mu$VK^{-1}) greater than that of pure films ($-57\,\mu$V K^{-1}), which increases directly proportional to temperature for both systems. In addition to greater S, the nanowires also have greater σ despite having lower carrier concentration than equivalently prepared PEDOT films because electron mobilities were reportedly greater in the nanowires by a factor of 3, resulting in PF as high as $92\,\mu$W m^{-1} K^{-2}. Gong *et al.* obtained well-defined crystalline nanowires of rigid rod conjugated polymer, a poly(paraphenylene ethynylene) derivative with thioacetate end groups, by the self-assembly process within a dilute solution.[93] Field-effect transistors utilizing the nanowires showed carrier mobilities (approaching 0.1 cm^2 V^{-1} s^{-1}) three to four orders greater than thin-film transistors utilizing the same polymer, indicating that the one-dimensional crystals have greater transport capability. Wang *et al.* fabricated well-defined microwires of CuPcOC$_4$ with controllable morphologies and diameters (about 1 μm, and lengths of 260–300 μm), as well as ultralong microwires (diameters of $\sim$10 μm, and lengths >5 mm), and σ is reported to be enhanced for the 1D microwires as compared to 2D thin films.[94] Regarding n-type polymers, Bertarelli and co-workers fabricated P(NDI2OD-T2) fibers (1–1.5 μm diameters) by electrospinning P(NDI2OD-T2)/poly(ethyleneoxide) blends followed by rinsing with a selective solvent, which results in preferential orientation of P(NDI2OD-T2) polymer chains along the fiber axis. Transistors consisting of a single fiber as the semiconductor were reported to have analogous electron mobilities (0.5–0.9 cm^2 V^{-1} s^{-1}) and on/off ratios compared to the best thin-film-based devices.

Significantly, thermal conductivity is consistently very low for polymeric systems ($<$1 Wm^{-1} K^{-1} at 300 K), which makes it worthwhile to try increasing σ by decreasing interchain separation, optimizing doping, and/or adding inorganic inclusions with high S and σ.

6.3 Morphological Considerations: Polymer–Inorganic Composites

Composites can be obtained by loading polymers with fillers, typically using micro- or nanoparticles, both pristine and functionalized, to create a conductive network within the more insulative matrix. Charge-transfer interactions between the fillers and the polymers increase the conductivity of the polymers by doping; this is a way to facilitate electrical conduction while dampening thermal vibrations. Polymer–filler heterojunctions utilize energy level off-sets and band structure to allow electrical transport and minimize thermal conductivities by carrier scattering or energy filtering at the interfaces.

For polymer-based inorganic hybrid composites, the formation of conducting pathways composed of both inorganic and organic constituents may be required because it is thought that S is "weighted" by the conductivity of individual paths. A homogeneous composite made of different components functioning independently cannot have a ZT higher than the highest value of any single component. In many cases, however, constituents can interact and may create internal potentials or interfaces with modified energy level distributions. For example, interfacial charge transport across hybrid composites has been controlled by matching the work function and band gap of different components to obtain high conductivity and S enhancements by energy filtering. Morphology also influences these parameters and the thermal conductivity via scattering mechanisms. Energy filtering can also be achieved via ionized impurity scattering.

In many cases, σ determines the PF in conducting polymers because either S is not changed much with increasing additive concentration or σ remains relatively low. Thus, understanding and increasing σ are thought to be the most promising routes for increasing ZT in polymer–inorganic composites, although enhancing S is promising because PF is proportional to S squared. Rational engineering of the organic–inorganic interfaces, including morphology and energy alignment, can allow improvement in PF by substantially enhancing S without greatly suppressing σ. The energy filtering effect may be achieved, even with microparticles, if composites are prepared such that percolation, or possibly near-percolation, of the particles is reached and only a thin layer (nano- or possibly even micrometers thick) of polymer coats the particles. Qiu *et al.* demonstrated the energy filtering effect in blended films of $FeCl_3$-doped P3HT and Bi_2Te_3 nanowires, where low-energy carriers were scattered more strongly than high-energy carriers by the interfacial potential barrier.[18] The composites utilizing greatest doping (32 wt% $FeCl_3$) and maximum Bi_2Te_3 loading (20 wt%) had greater TE performance (PF $\sim 13.6\,\mu W\,m^{-1}\,K^{-2}$, $S \sim 128\,\mu V\,K^{-1}$, $\sigma \sim 300\,S\,m^{-1}$, $\kappa \sim 0.86\,W\,m^{-1}\,K^{-1}$) than just 32 wt% $FeCl_3$ polymer films (PF $\sim 3.9\,\mu W\,m^{-1}\,K^{-2}$, $S \sim 55\,\mu V\,K^{-1}$, $\sigma \sim 1{,}800\,S\,m^{-1}$, $\kappa \sim 0.54\,W\,m^{-1}\,K^{-1}$). Interfacial effects and design of energy levels, or density of states for different carriers, are the critical factors to consider for improving systems that already show promise.

Energy level offset between semiconductors and ohmic contacts with electrodes are critical to consider. Zhang *et al.* identified contact resistance between constituents as the TE performance limiting factor in composites of PEDOT blended with bismuth telluride particles, which were not synthesized or grown within the polymer.[7] Oxidation of inorganic particles during synthesis or handling may cause interface resistances, which can be avoided by *in situ* formation of the inorganic during processing. However, the redox potentials and dopant stability will also be critical to consider so that inorganic particles are not oxidized during *in situ* polymerization either. Unfortunately, interfacial regions between the inorganic and organic phases are not well understood, or the nature of bonding between them, although some groups have investigated this aspect in particular.[18,20,101] Ohmic connections and low-energy barriers have been shown to increase the power factor of composites, but charge transport through composites remains theoretically complicated and difficult to map experimentally. Most likely, there is an optimal particle size and distribution depending on both the thermal and electrical energy alignment and charge transfer between constituents. It is also critical to consider effects at all length scales from atomic to mesoscopic.[102]

It might seem that less surface area contact between polymer and particles is desirable to reduce contact resistances and voltage drop losses or barriers, but this may not be true if the density of states and Fermi energy are designed properly such that charges transport through the material, but not heat/phonons. If the inclusions are also used as dopants, then more surface area contact between the dopant and the polymer is desirable to enhance carrier concentration, and possibly for creating more energy states that enhance S. In this case smaller particles with great homogeneous dispersion are favorable for increased interfacial connection. Still, smaller dopants (i.e., nanoparticles and molecules) have the tendency to diffuse in organic systems, unlike macroscopic crystals or large molecules, leading to more detrimental instability.[99,100]

Phonon scattering, typically achieved by increased surface/interfacial area due to reduction in particle size, is less necessary in polymer composites as it is for inorganic composites due to intrinsically lower κ. Nanoinclusions, as opposed to microinclusions, may be less necessary in the case of TE polymer composites because the thermal conductivity of composites with high additive concentrations still correlates strongly in most cases with the thermal conductivity of the bulk polymer. However, highly loaded polymers (i.e., when polymer content is low, <40 wt%) consistently show the best TE results because inorganic TE components generally have higher TE performance, they assist in forming long-range transport pathways, and because transport efficiency is limited by lower mobility carriers.[8,11,16,96,145–149] Increasing the particle/polymer weight ratio to an optimal loading has been shown to increase transport efficiency between electrodes, which is attributed to the percolating network formed by the crystalline particle phase. Optimal weight ratios of particle:polymer are larger for standard polymers without functional groups specific to the inorganic inclusions (9:1),[145–149] compared to functionalized and chemically interactive constituents (1:1).[150] Cao *et al.* indicate that for bare spherical nanocrystals (5 nm diameter) of CdSe in poly[2-methoxy-5-(2′-ethylhexyloxy)-*p*-phenylene vinylene] that an efficient interpenetrating network of inorganic crystallites is formed by increasing the CdSe concentration from 83 to 90 or 94 wt%.[96] The efficiency of transport can also be improved by utilizing anisotropic particles and structures, which is attributed to enhanced percolation,[151–155] and also due to longer carrier lifetimes.[156]

Structures with higher aspect ratios can form conductive networks at relatively lower volume fractions, which are preferred for applications because they would require less material and inorganics typically have higher density than polymers, which would increase the weight of the composite. Significantly, nanostructuring increases the surface area to volume ratio greatly at the cost of greater series contact resistances. Although microcomposites are generally less homogeneous compared to nanocomposites, they may have greater conductivity and less scattering surface area because larger inclusions have greater carrier mobilities due to longer mean-free path distances. However, larger particles could also contribute greater thermal conductivity, although this has not been shown in the TE literature due to difficulty of thermal conductivity measurements and lack of understanding in regard to electrical–thermal coupling in various conducting polymers.[16] The thermal conductivity of conducting polymers is insensitive to dopant concentration in many cases,[36,80,97,98,110] but not always.[104–107]

Ultimately, preparation of micro- and nanocomposites from unmodified components is not trivial, and simple casting from solvent often leads to phase segregation of constituents at the micrometer scale.[157] Purely physical methods have been reported in attempts to controllably distribute individual nanocrystals (which are not modified, but may be capped with inert ligands) within a polymer matrix (which is also not functionalized). One method includes exposing the polymer (P3HT) casting zone to an electric field, leading to an oriented supramolecular texture due to phase segregation and alignment of nanocrystals (CdSe), which is particularly viable when crystals exhibit higher aspect ratios.[158] Nanocrystals capped with initial ligands, such as oleic acid, stearic acid, hexadecylamine, and tri-octylphosphine oxide, have natural tendency to form densely packed agglomerates (ordered supercrystals) upon solvent removal.[159,160] Another route is directional epitaxial solidification of the polymer (P3HT) in a crystallizable solvent (1,3,5-trichlorobenzene) in the presence of spherical or rod-like nanocrystals (CdSe).[161] Solidification leads to semicrystalline polymer domains and lamellar structure alternating with amorphous interlamellar layers, and nanocrystals are rejected to amorphous polymer regions. Conjugated polymers exhibit varying semicrystalline structures and morphologies upon solidification using appropriate solvents and temperature regimes, which leads to a range in performance.[52–55,80–85,89–95,162,163] Xu *et al.* demonstrated that solidification of P3HT in a high boiling point solvent (*p*-xylene or cyclohexane) leads to nanofiber formation (20 nm diameters).[164] CdSe nanocrystals self-assemble on the surface of nanofibers, resulting in enhanced photoconductivity compared to polymer films or composite films fabricated by blending constituents and removal of solvent by evaporation at room temperature. Although strongly binding ligands to particles enhances solubility compared to loosely bound ones, strongly bound ligands have been shown to inhibit dispersion within the bulk composite and electronic coupling (i.e., charge transfer) between polymer and particles.[146] Chemical methods for obtaining controllable crystalline dispersion include capping nanocrystals with conjugated macromolecules,[150,165–169] covalent grafting of functionalized nanocrystals with the conjugated oligomer or polymer,[170–174] noncovalent grafting interactions,[175–179] and *in situ* growth of nanocrystals within conjugated polymer matrices.[180–185]

6.4 Polymer–Inorganic Composites: p-Type Devices

Progress toward greater power factors has been achieved in p-type polymers by incorporating high-performance inorganic TE materials, or by doping with a sufficient amount of suitable dopants. In these cases the inorganic particles are electron trapping, which assist hole conduction within the p-type polymer composite. However, an increase in dopants (increase in conductivity) can lead to the expected decrease in the S coefficient.[103–105] In that case, the number of charge carriers increases and the Fermi energy is forced toward the conduction band.[106] Some examples of doping agents are metal halides such as iron(III) chloride ($FeCl_3$),[103] and iodine,[107] and metal oxides such as molybdenum trioxide (MoO_3, Fig. 6.3(d)).[108] Consideration of the energy difference between the HOMO level of the polymer and the Fermi energy level of the dopant is critical for efficient doping, although it has been shown that more homogeneous dispersion can be more effective in raising carrier concentration, like the example of fluorinated molybdenum complex dispersed in p-type 1,4-bis[N-(1-naphthyl)-N'-phenylamino]-4,4'-diamine (NPB).[109] High-performance inorganic TE materials (such as tellurides) are the most promising for achieving high S in polymer–inorganic composites. Carbon nanostructures can have significantly different effects,[110–114] which will be discussed briefly.

There have been many studies of polymer–inorganic composites. The most promising are based on PANI, PEDOT, and PTH, although other polymers are being explored (PPY, carbazole-based polymers) and new polymers are emerging (imide-based polymers). Composites of PANI and inorganic compounds were the first polymer–inorganic TE systems to be studied (1990s), and are still being investigated today.[1–4,6,9,10,13,31] PANI has been blended with metal oxides,[1] Bi,[4,6] tellurides,[2,9,10,13,31] $NaFe_4P_{12}$,[3] and carbon nanotubes.[110,111]

Recently, the dispersion of Te nanostructures in an aqueous solution of PEDOT was obtained and cast to form composites with ZT of about 0.1 at room temperature (Fig. 6.2(a)).[8,19,20] Due to the S coefficient being positive and significantly higher than pristine polymer, it is thought that the holes were solely responsible for charge transport and that the transport did not occur exclusively through the polymer. The use of Te nanostructures in thermoelectrics was suggested earlier, and the fundamental thermoelectric properties of elemental Te have appeared in the literature extending decades back.[115,116]

Polymer-assisted growth of particles *in situ* of thin film fabrication has been shown to improve interfacial interactions, including electronically driven interactions between polymer and as-grown particles. *In situ* growth of the metal-based particles within the polymer matrix is also expected to enhance the interactions between constituents and can lead to a charge complex at the interface. One example is the interaction between Te nanowires grown in PEDOT.[19,20] Urban *et al.* suggest that carrier transport occurs predominantly through a more conductive volume of PEDOT that exists at the interface with tellurium nanocrystals, which may be due to good morphological structure accomplished by *in situ* fabrication.[4] Similarly, cadmium telluride nanocrystals were synthesized in P3HT without use of surfactants.[101] Ultraviolet–visible absorption and photoluminescence quenching studies reveal that the nanocrystals work as transport media, which are bound to polymer via dipole–dipole interactions and form a charge transfer complex, facilitating percolation pathways for charge transport. Structural/morphological studies reveal that the nanocrystals work as transport media, which facilitate percolation pathways for charge transport, although an increase in the carrier concentration (and/or trap density for hole transport) appears to be most responsible for enhancing σ.

Segregated-network carbon-nanotube (CNT) PANI composites have significantly increased σ when percolated CNT networks were obtained.[110–112] S and σ both increase in the case of SWNTs because the pi–pi interactions create a more ordered structure and generate higher carrier mobility.[110] Thermal conductivity remains constant at high CNT loadings, up to 41.4 wt%. Another group obtained significantly improved PF (0.5–$5\,\mu\text{W}\,\text{m}^{-1}\,\text{K}^{-2}$ at $300\,\text{K}$) using different processing, which is likely due to the size-dependent energy-filtering effect caused by thin nanostructured PANI coating between CNTs.

P3HT/MWCNT composites were prepared by ball milling, so no growth or formation of nanotubes occurs during polymer formation other than agglomeration of particles.[113] The κ (0.1–$0.8\,\text{W}\,\text{m}^{-1}\,\text{K}^{-1}$), σ (2–$6\,\text{S}\,\text{cm}^{-1}$), and S (10–$17\,\mu\text{V}\,\text{K}^{-1}$) all increase with the increase in the concentration of MWCNTs. The ZT only reaches 0.0002 at elevated temperature. The TE properties, S and σ, were also enhanced simultaneously in the case of doped PANI nanotubes,[52] stretched PANI films,[55] and urea-doped PEDOT:PSS.[81]

Hierarchical fullerene-graphene nanohybrids were successfully synthesized. TEM, XRD, Raman, and UV–vis spectra confirmed the decoration of fullerene on graphene, and charge transfer between them.[114] Thermoelectric materials were fabricated by integrating such hierarchical nanohybrids into PEDOT:PSS. In the hierarchical nanohybrid-filled polymer composites, fourfold improvement in the S coefficient was achieved as compared to the neat polymer film due to the potential interfacial energy filtering. Graphene/polymer composites and fullerene/polymer composites were also investigated, but their ZT was around 10^{-3}. It was shown that graphene and C_{60} can balance the conflicts of the thermal/electric transport and together resulted in the largest ZT. Tuning the ratio of C_{60} to graphene in the nanohybrids can make the σ increment surpass the increase in the thermal conductivity, resulting in an optimized ZT around 0.07, which is more than one order of magnitude improvement compared to the single-phase filler-based polymer composites.

Kim *et al.* found that PEDOT:PSS particles are spread on the surface of CNTs in their composites, bridging CNT–CNT junctions and allowing electrons to travel through the composite and thus increasing σ.[117] Significantly, heat transport in this system is suppressed due to differences in vibrational spectra between CNT and PEDOT:PSS. The highest achieved ZT was 0.02 when 35 wt% CNT was added to the polymer matrix. The addition of CNT as a dopant beyond a certain threshold also resulted in the decrease of both electrical and S coefficient,[118,119] and an increase in the thermal conductivity.[112]

Hu *et al.* combine common metals (i.e., Al, Au, Cu) and p-type acid-doped PPY into hybrid metal/polymer/metal trilayer composites with submicron thicknesses (similar to Fig. 6.1(b)).[39] The metal electrodes used to sandwich the conducting polymer also serve as hot and cold TE electrodes, as usual, but in the sandwich configuration the thermal transport is through the plane of the composite, as opposed to in-plane for laterally defined structures. The vertical structure presumably creates a bottleneck to thermal conductivity and an enlarged entropy difference in the internal energy of charge carriers between high- and low-temperature metal electrodes due to the sharp change in properties over a large capacitor-like area, leading to larger S. The thin metal/polymer/metal films also exhibit high σ due to thin spun-cast polymer layers, about 200 nm thick, particularly for doped PPY (as opposed to undoped) which forms ohmic contacts with Al and Au.

At room temperature, Al/PPY/Al devices with polymer thicknesses of 170 nm and 230 nm both exhibited S around $100\,\mu\,V\,K^{-1}$. The conductivity was reported to be $3.5 \times 10^{-6}\,S\,cm^{-1}$ for the doping level of the 1:3 molar ratio between dopant and PPY monomer unit, but was not reported as a function of temperature. PPY devices with varying polymer conductivities (by changing dopant concentration) were not investigated. S increases with temperature for both thicknesses, but by apparently different functions of temperature, and the thicker film reaches $200\,V\,K^{-1}$ at 70°C, while the thinner film barely reaches $150\,\mu\,V\,K^{-1}$. Presumably at elevated temperatures the two functions could cross and the thinner film may exhibit greater S. This was demonstrated by Hu *et al.* when they made Al/PEDOT:PSS/Al and tested them at 30–110°C.[40] The thicker polymer film (300 nm) shows a stronger initial increase in S with temperature, and the thinner film (200 nm) starts off with lower S but is more sensitive to a sharp increase in S brought about by elevated temperatures. Both Al/PEDOT:PSS/Al devices start around $20\,\mu\,V\,K^{-1}$, then cross at about 85°C with $100\,\mu\,V\,K^{-1}$, and the thin film reaches $250\,\mu\,V\,K^{-1}$, while the thicker film reaches $150\,\mu\,V\,K^{-1}$.

Temperature-dependent S suggests that increasing temperature can largely boost charge density but still limits thermal conduction through interfacial phonon scattering. As a consequence, high S is obtained at high temperatures. In addition, varying the polymer film thickness can vary S such that thinner films allow better electrical conduction due to the thinner low-conductivity polymer pathway, and have greater capacitance, leading to lower S at lower temperatures but greater S compared to thicker films at high temperature. The metal/polymer interface can limit thermal conductivity by acoustic mismatch,[40] phonon scattering at defects and boundaries,[41,42] and low phonon density of the polymer which makes it a barrier to thermal conduction.

Hu *et al.* also fabricated the Al/PPY/Au(cold side) and Au/PPY/Al (cold side) devices, where the Au top contact contributed to greater σ (6.8×10^{-6} compared to $8.1 \times 10^{-7}\,\mathrm{S\,cm^{-1}}$) due to relative penetration of different metals from thermal evaporation. It was demonstrated that devices with greater σ also had greater S coefficients (200 compared to $150\,\mu\mathrm{V\,K^{-1}}$ for Al/PPY/Au(cold side) and Au/PPY/Al(cold side) measured at 70°C), although the Al/PPY/Al device has exactly the same S (over the same range of temperatures) and lower σ.

Hu *et al.* also fabricate semiconducting photovoltaic bilayer systems comprised of Au/PEDOT/P3HT-blend/Al, which function as both thermoelectric ($S = 20\,\mu\mathrm{V\,K^{-1}}$) and photovoltaic devices if the Al is semitransparent (2.26% photoconversion efficiency at 1 sun).[39] Depending on the concentration of dopant phenyl-C61-butyric acid methyl ester (PCBM, Fig. 6.3(o)), the P3HT blend can behave more n-type or p-type. Using an ITO/polymer/metal device under a light intensity of $16\,\mathrm{mW\,cm^{-2}}$, Hu *et al.* show that photoexcitation can lead to S enhancement (increased from $100\,\mu\mathrm{V\,K^{-1}}$ to $300\,\mu\mathrm{V\,K^{-1}}$).[41] Simultaneously σ increased in the light to $8.2 \times 10^{-5}\,\mathrm{S\,cm^{-1}}$ from $3.6 \times 10^{-6}\,\mathrm{S\,cm^{-1}}$ in the dark, showing that the excited states can lead to an increase in PF. Experimental studies show that photogenerated excited states increase the entropy difference between hot and cold surfaces through electron–phonon coupling and increased charge carrier density through dissociation, which negligibly affect thermal transport (device structure seen in Fig. 6.2(d)).

Fibers can be reliable scaffold materials for functional polymers, and have been applied to TE applications (similar to Figs. 6.2(f), or 6.2(c) and 6.2(e)). There have been numerous reports on the deposition of air-stable conducting organic polymers such as PPY, PANI, or PEDOT onto organic or inorganic fibrous substrates.[120–125] Conducting polymer-coated fibers has attracted much interest due to the increasing applications of these fibers for microwave attenuation, static charge dissipation, and electromagnetic interference shielding. While imparting σ, the polymer coating process barely affects the strength, drape, flexibility, and porosity of the starting substrate. Single fiber devices are typically tested in thermocouple geometry using a reference metal wire as opposing conductor.

Fiber structures could also be woven into lightweight, high-strength, multifunctional textiles for seamless integration with structural composites. The fiber form factor is a powerful paradigm for these energy conversion devices, since it can allow a high density of thermocouple junctions without the use of costly patterning techniques. Conversion of heat to electricity is accomplished by the conventional series-connected junction geometry, which is reproduced in the form of thin-film segments deposited concentrically around a fiber core (Fig. 6.2(f)). Weaving these fibers appropriately can position the junctions as required for power generation. Woven thermoelectric generators have been demonstrated utilizing several fiber diameters and also varying the TE segment length and weave density.[33,34,126,127] The power density increases dramatically for smaller fibers, higher weave density, and higher temperature gradients. The thinness and flexibility of these mats suggest that multilayer TE fabrics can be used to efficiently span temperature gradients using individual layers tuned to work at their maximum ZT. Currently there are great efforts to commercialize electronic textile devices, and it is likely that TEGs (and batteries, displays, etc.) will be incorporated within textiles as a standard procedure in the near future.

Li *et al.* fabricated PPY-coated fabrics (textile substrate was not specified) to test the hypothesis whether the thermoelectric effect occurs if a DC current is applied across two dissimilar conducting polymers, for which S and Peltier effects were indeed observed, although they were unsteady primarily due to degradation of the conducting polymer.[33] The dissimilar PPY fabrics were prepared by different methods, and a thermocouple junction between the two conductors, placed across a temperature difference, was used in order to obtain an S effect ($\sim 10\,\mu\mathrm{V\,K^{-1}}$). One PPY fabric was obtained by immersing the textile in aqueous $2.4\,\mathrm{mg\,ml^{-1}}$ PPY solution with $14\,\mathrm{mg\,ml^{-1}}$ ferric chloride, and the other was prepared by oxidation from immersion in $10\,\mathrm{g\,l^{-1}}$ ferric chloride followed by exposure to pyrrole monomer vapor. Significantly, the S of the polymer-coated fiber thermocouples is comparable to metals and much lower than the bulk conducting polymer ($>100\,\mu\mathrm{V\,K^{-1}}$). A thermocouple of the PPY-coated fabric and aluminum obtained the S effect of $150\,\mu\mathrm{V\,K^{-1}}$.

Henn *et al.* studied the TE properties of commercially available PPY-coated polyester fibers (anthraquinone-2-sulfonic acid is used as PPY dopant in 0.33:1 molar ratio).[34] The S of PPY-coated textiles is around $5\,\mu\mathrm{V\,K^{-1}}$, obtained by placing PPY fabric electrically in the thermocouple junction with copper wire. The PF of PPY-coated fibers is about $2\,\mu\mathrm{W\,m^{-1}\,K^{-2}}$ ($5\,\mu\mathrm{W\,m^{-1}\,K^{-2}}$ nonwoven) and ZT = 0.002 (0.005 nonwoven). Nonwoven refers to single polyester fiber coated in PPY compared to a fabric of woven fibers. Hollow fibers made completely of PPY, and PANI, using a sacrificial PMMA, and poly(L-lactide), fiber template have been demonstrated, with diameters less than $1\,\mu m$.[125,125] The σ of the PANI tubes was slightly reduced ($0.28\,\mathrm{S\,cm^{-1}}$) compared to the scaffold PANI fiber ($0.38\,\mathrm{S\,cm^{-1}}$).

In comparison to these organic polymer–fiber composites, Shtein *et al.* fabricated flexible TEGs by thermally evaporating inorganic semiconductor thermocouple junctions (specifically Bi_2Te_3 and Sb_2Te_3) or metal thermocouple junctions (silver and nickel) onto flexible silica fiber substrates. Telluride thermocouples achieve an open circuit S of about $375\,\mu\mathrm{V\,K^{-1}}$ and a predicted power output of $1\,\mu\mathrm{W}$ per Bi_2Te_3–Sb_2Te_3 thermocouple for the segment length of $5\,\mathrm{mm}$ using a hot junction at $370\,\mathrm{K}$ (Fig. 6.2(f)).[126] Having the same coated layer thickness, Ag–Ni devices exhibit an S of $20\,\mu\mathrm{V\,K^{-1}}$, so the generated power is predicted to be five times less. Polyimide fibers lead to higher theoretical power output due to lower thermal conductivity of the substrate, which allows a greater temperature difference to be maintained per distance, so shorter segment lengths become optimal. Fiber-based designs utilizing hybrid materials, organic conducting polymers, and inorganic semiconducting particles have yet to be realized.

6.5 Polymer–Inorganic Composites: n-Type Devices and Complete Modules

As expected, the previously described TE composites all behave p-type, exhibiting positive S coefficients. Viable thermoelectric devices require both electron- and hole-conducting elements. However, there are few examples of stable extrinsic n-type doping of organics, and even fewer studies of their thermoelectric properties. Extrinsic n-type doping of organic semiconductors presents a challenge due to their small electron affinities (-3 eV to -4 eV). Polymer–inorganic composites utilizing unipolar n-type conducting polymer have not been reported. Only vapor-deposited doped fullerenes[65–68] and the organometallic poly(Kx(Ni-1,1,2,2-ethenetetrathiolate))[38] demonstrate high room temperature PF, around 30–70 μW m^{-1} K^{-2}. In addition to their insolubility, these materials still have TE values lower than PEDOT. Importantly, some composites made entirely of hole-transport materials can have n-type S coefficients, due to an effect of the processing on the composition and final structure of the material resulting in a change in the energy landscape, presumably doping the polymer, or greatly increasing the concentration of traps to positive carriers, or filling traps to negative carriers.

Hydrochloric acid (HCl)-doped PANI exhibited a negative S coefficient (from $-6\,\mu$V K^{-1} to $-93\,\mu$V K^{-1}), which increases with temperature (303–423 K) as σ decreases (S cm^{-1}) over the same T range.[97] Increasing the dopant concentration of HCL-doped PANI results in ZT initially increasing and then decreasing (maximum $\sim$2.7 $\times$ 10^{-4}), which follows behavior of κ (0.28 W m^{-1} K^{-1} to 0.46 W m^{-1} K^{-1} and back down to 0.34 over the T range). Ultralong PEDOT nanowires exhibit negative S around $-122\,\mu$V K^{-1} due to greater electron mobility or carrier concentration, presumably due to electropolymerization in aqueous LiClO$_4$ solution.[93] Also, an S of $-1{,}000$ to $-4{,}000\,\mu$V K^{-1} has been observed in PEDOT nanotubes that were modified with lead telluride (p-type) and then sintered into macroscopic pellets, but PF reaches only 1 μW m^{-1} K^{-2}.[12] Also, n-type PTH-Bi$_2$Te$_3$ was prepared by hydrothermal synthesis of Bi$_2$Te$_3$ nanopowder (p-type) and oxidative polymerization to obtain polymer, followed by pressing (80 MPa) the mixture into pellets at different temperatures.[11] The highest S ($-98\,\mu$V K^{-1}) was obtained at 473 K, while the highest power factor (2.54 W m^{-1} K^{-2}) obtained at 623 K was due to highest σ (8 S cm^{-1}), and was 20 times larger than that at 473 K, which are still lower than pure Bi$_2$Te$_3$ compressed at the same temperature. Composites can display switch in sign of S at a critical temperature or dopant concentration, most likely due to the shift of the Fermi level. For instance, PPY samples that have been aged (held 24–120 days at 353 K) show a change in sign to negative thermopowers at low temperature (below 200 K depending on aging time and material).[75,79]

Zhang *et al.* measured a PF of $80\,\mu\mathrm{W\,m^{-1}\,K^{-2}}$ for an n-type Bi_2Te_3 particle–PEDOT composite.[7] Bi_2Te_3 micropowder was blended with PEDOT by ball milling the Bi_2Te_3 and mixing it in solution with the polymer. A two- to threefold increase in power factor relative to either constituent alone was obtained. Significantly, they demonstrated that n-type TE polymer composites could be obtained by loading high-conductivity and high-S n-type powders into a p-type conducting polymer with low S (achieved 60–$80\,\mu\mathrm{W\,m^{-1}\,K^{-2}}$ using 70–90 vol% inorganic powder, approaching $ZT = 0.1$). The contact resistance between inorganic and polymer is likely the limiting factor for further TE property improvement, requiring further interfacial engineering. An n-channel host material would potentially be better by not canceling S coefficients. In addition, Bi_2Te_3, like many other inorganic thermoelectric materials, is undesirably composed of less common and more toxic elements.

Lu *et al.* fabricated organic TE devices with both p-type (P3HT:PCBM but P3HT rich) and n-type (P3HT:PCBM but PCBM rich) conducting polymers and compared their combined performance (p–n module) with p-type-only or n-type-only devices.[37] The p-type devices are based on P3HT with concentrations of 10, 20, and 30 mol% PCBM. N-type devices are PCBM composites with 10, 20, and 30 mol% P3HT. As expected, but finally proven for polymer-based thermoelectronics, the combined module greatly enhanced the performance due to the separated charge carrier transport channels for holes and electrons. In order to maintain S while doping the P3HT polymer, they used photoexcitation to counteract the S reduction by generating an additional concentration of minority carriers with strong phonon–electron coupling, which also increases conductivity. The combination of improvements yields a power factor of around $400\,\mu\mathrm{W\,m^{-1}\,K^{-2}}$ and a ZT of about 0.5 at 150°C. This ZT value is comparable to inorganic superlattice devices utilizing

Bi or Sb tellurides, deposited by thermal evaporation or printed from suspension.[128–133]

For comparison, Evans *et al.* use inorganic-epoxy solutions for ink-jet style printing and demonstrate a planar printed TE device which can be rolled up, but the performance is comparable to all-organic printed devices.[129,130] The device is operated in the rolled-up configuration because the film produces a voltage in-plane along its surface, and rolling eventually produces a large surface as hot and cold side for through-plane operation. Cho *et al.* fabricate flexible inorganic devices by printing n-type Bi_2Te_3 and p-type Sb_2Te_3 onto glass fiber fabrics.[131] Devices are thin and light, about $500\,\mu$m thick and about $0.13\,$g cm^{-2}. Output power per unit length is $28\,$mW g^{-1} at $50\,$K. Lu *et al.* fabricate devices by printing suspensions of nanoparticles made of ternary telluride alloys.[132] Eason fabricated devices with Bi_2Te_3, Bi_2Se_3, and $Bi_{0.5}Sb_{1.5}Te_3$ via laser-induced forward transfer of intact solid films using a nanosecond excimer laser.[133] The resulting S and series resistance per TEG leg pair was $170\,\mu$V K^{-1} and $10\,$KΩ, respectively.

Fujifilm developed ink-jet printed TEGs using PEDOT:PSS. The exhibited thermoelectric converter module has a power generation capacity of several milliwatts ($ZT > 0.27$) and is capable of generating electricity with a temperature difference of $1°$C.[145] In order to achieve the necessary depth, TEGs require to transform temperature differences into electric power; the currently available 3D printing process could be very well suited. 3D printing technology works like an inkjet printer, but a thermoelectrically active polymer paste is printed instead of a thin ink, and many layers can be built up resulting in films deposited up to $30\,\mu$m thick for devices. Fraunhofer Institute for Material and Beam Technology IWS recently developed 3D-printed TEGs using conductive polymers for low temperature applications, up to $200°$C.[146]

Carbon fiber composites have been demonstrated as a cooling fabric with n-type S (Fig. 6.2(g)).[35] Pickering *et al.* measure the σ and S of intercalated recycled carbon fiber sheets derived from polyacrylonitrile precursor. Significantly, S and σ are found to be unchanged by the repeated fluidized bed recycling process; S

increases almost twofold (from $-9 \, \mathrm{V \, K^{-1}}$ to $-19 \, \mathrm{V \, K^{-1}}$) upon inter-calation with sulfuric or nitric acid, while σ increases through n-doping by a factor of 1 from its initial value (no explicit value is reported). The highest S obtained is comparable to bulk nickel at room temperature, but the σ of intercalated carbon fibers increases over a range of increasing temperature, rather than decreasing.

Carroll *et al.* designed a multilayer CNT/polymer film that allows for the arrangement of the temperature gradient parallel to the film's surface (Fig. 6.2(e)).[42] The film consists of alternating layers of conducting material (polyvinylidene fluoride (PVDF) polymer containing multiwalled CNTs) and insulating material (pure PVDF) bonded together by heat and pressing. Each layer has a thickness of 20–40 μm. The generated thermoelectric voltage is equal to the sum of contributions from each layer, so adding layers to the fabric is equivalent to adding voltage sources in series, and the number of layers is limited only by the heat source's ability to produce a sufficient change in temperature throughout all the layers. Experiments on a 72-layer fabric demonstrated a maximum power generation of 137 nW at a temperature difference of 50 K. But the researchers predict that the power output can be increased; for example, they calculate that a 300-layer fabric exposed to a 100 K temperature difference has a theoretical power output of up to 5 μW. The fabric's layers add somewhat linearly, which allows production in a form factor that promotes large area applications.

C_{60} is not a polymer, or amenable to solution processing, but it has been utilized like a polymer matrix for inorganic inclusions, or as an inclusion within inorganic matrices. This unique molecule has high electron affinity, and can contribute to cooperative phenomena, such as superconductivity and ferromagnetism when doped or in complex with other molecules.[134–136] Gothard *et al.* demonstrated fullerenes as a process control agent.[137] C_{60} can engage in significant microstructural refinement, even in small quantities, of composites based on bismuth telluride. The Bi_2Te_3/C_{60} composites are synthesized via ball milling using different concentrations, followed by spark plasma sintering into pellets. C_{60} actively refines Bi_2Te_3 particle size during milling by inhibiting formation of larger particles.

Fullerene-assisted refinement causes a decrease in lattice thermal conductivity and σ, but increases S (from $70\,\mu\mathrm{V\,K}^{-1}$ to $200\,\mu\mathrm{V\,K}^{-1}$), which results in an enhancement of the PF at low volume fractions. Blank *et al.* show that C_{60} decorates Bi–Sb–Te nanocrystals when synthesized by planetary milling, and the fullerenes act as electron traps and phonon scattering sites.[138] For their p-type Bi–Sb–Te/C_{60} composites, ZT = 1.15, which is 30% greater than the undoped Bi–Sb–Te system, although Bi_2Te_3 bulk material can have ZT from 1 to 1.5, and in that report they also reference a Bi–Sb–Te alloy produced by ball milling and sintering having ZT = 1.4.[147] Skudderite materials have also been modified by the addition of C_{60}, which forms microclusters between grain boundaries.[139] Also, Cs_2CO_3/C_{60} composites are investigated in a bilayer configuration comprising vertically stacked thin films.[67] Utilizing C_{60} films with thicknesses between 20 and 50 nm as an underlying layer improved the PF compared to pure Cs_2CO_3 films (5 nm thickness) by an order magnitude, resulting in a maximum PF of around $20\,\mu\mathrm{W\,m}^{-1}\,\mathrm{K}^{-2}$.

Dang *et al.* dope C_{60} films using acridine orange base, an organic molecule which reduces the fullerene film and increases carrier concentration.[140] Riede *et al.* achieve up to $4\,\mathrm{S\,cm}^{-1}$ by incorporating Cr_2- and W_2-based dopants into C_{60} films. The activation energy of conductivity and S decrease upon increasing doping concentration, indicating a shift of Fermi level toward the electron transport level of C_{60}.[68] Bao *et al.* synthesized a small tungsten-based molecule as dopant for C_{60}, showing a conductivity of $5.5\,\mathrm{S\,cm}^{-1}$ at 8 wt%, the highest yet reported for molecular n-type semiconductors.[141] Conservation of n-doping after exposure to air is investigated by conductivity, ultraviolet photoemission spectroscopy, and S measurement. One third of the initial conductivity can be restored after dropping fivefold in air by annealing in vacuum, explained by self-passivation of the molecular n-doping. Protection against oxidation in air may be due to down-shift of energy levels of unstable dopant upon charge transfer to a host material with deeper lying energy levels. The high electron affinity makes C_{60} an efficient p-type dopant in conjugated polymers, essentially by trapping electrons. However, the LUMO of C_{60} lies between the top of the valence band and the bottom of the conduction band in most polymers, indicating weak doping ability,[142] although photo-induced charge separation can lead to photoconductivity comparable to highest performing photoconductors.[143,144]

6.6 Conclusions

Hybrid composites involving polymers and inorganic particles are complex, having properties that depend on molecular and macromolecular parameters of the polymer and particle constituents in unison. Polymer chain microstructure, molecular mass, and the distribution of properties over various scales must be considered, as well as the shape, size, distribution, orientation, and stoichiometry of nanocrystals, in addition to the chemical and physical interaction between components. Highly ordered anisotropic nanocomposites using appropriately engineered semiconductors and interfaces will lead to the best TE performance.

References

1. C. G. Wu, D. C. DeGroot, H. O. Marcy, J. L. Schindler, C. R. Kannewurf, Y. J. Liu, W. Hirpo, and M. G. Kanatzidis, Redox intercalative polymerization of analine in V_2O_5 xerogel. The postintercalative intralamellar polymer growth in polyanaline/metal oxide nanocomposites is facilitated by molecular oxygen. *Chem. Mater.*, **8**, 1992 (1996).
2. X. B. Zhao, S. H. Hu, M. J. Zhao, and T. J. Zhu, Thermoelectric properties of $Bi_{0.5}Sb_{1.5}Te_3$/polyanaline hybrids prepared by mechanical blending. *Mater. Lett.*, **52**, 147 (2002).
3. H. Liu, J. Y. Wang, X. B. Hu, R. I. Boughton, S. R. Zhao, Q. Li, and M. H. Jiang, Structure and electronic transport properties of polyaniline/NaFe composite. *Chem. Phys. Lett.*, **352**, 185 (2002).
4. S. R. Hostler, P. Kaul, K. Day, V. Qu, C. Cullen, A. R. Abramson, X. F. Qiu, and C. Burda, Thermal and electrical characterization of nanocomposites for thermoelectrics. *Thermal and Electrical Characterization of Nanocomposites for Thermoelectrics ITHERM '06 10th Intersoc. Conf*, pp. 1400–1405 (2006).
5. E. Pinter, Z. A. Fekete, O. Berkesi, P. Makra, A. Patzko, and C. Visy, Characterization of poly(3-octylthiophene)/silver nanocomposites prepared by solution doping. *J. Phys. Chem. C*, **111**, 11872 (2007).

6. H. Ann, M. Fukamoto, Y. Heta, K. Koga, H. Itahara, R. Asahi, R. Sato-mura, M. Sannomiya, and N. Toshima, Preparation of conducting polyaniline–bismuth nanoparticle composites by planetary ball milling. *J. Electron. Mater.*, **38**, 1443 (2009).
7. B. Zhang, J. Sun, H. E. Katz, F. Fang, and R .L. Opila. Promising thermoelectric properties of commercial PEDOT:PSS materials and their Bi_2Te_3 powder composites. *ACS Appl. Mater. Interf.*, **2**, 3170 (2010).
8. K. C. See, J. P. Feser, C. E. Chen, A. Majumdar, J. J. Urban, and R. A. Segalman, Water-processable polymer–nanocrystal hybrids for thermoelectrics. *Nano Letters*, **10**, 4664 (2010).
9. N. Toshima, M. Imai, and S. Ichikawa. Organic–inorgnic nanohybrids as novel thermoeelctric materials: hybrids of polyanaline and bismuth(III) telluride nanoparticle. *J. Electron. Mater.*, **40**, 898 (2011).
10. Y. Li, Q. Zhao Y. G. Wang, and K. Bi. Synthesis and characterization of BiTe/polyaniline composites. *Mater. Sci. Semicond. Process*, **14**, 219 (2011).
11. Y. Du, K. F. Cai, Z. Qin, S. Z. Shen, and P. S. Casey, Preparation and thermoelectric properties of Bi_2Te_3/polythiophene nanocomposite materials. Conference on Mechanical, Industrial, and Manufacturing Engineering, pp. 462–465 (2011).
12. Y. Y. Wang, K. F. Cai, and X. Yao, Facile fabrication and thermoelectric of PbTe-modified poly(3,4-ethylenedioxythiophene) nanotubes. *ACS Appl. Mater. Interfaces*, **3**, 1163 (2011).
13. Y. Y. Wang, K. F. Cai, J. L . Yin, B. J. An, Y. Du, and X. Yao, In situ fabrication and thermoelectric properties of PbTe–polyanaline composite nanostructures. *J. Nanopart. Res.*, **13**, 533 (2011).
14. Y. Y. Wang, K. F. Cai, and X. Yao, One-pot fabrication and enhanced thermoelectric properties of poly(3,4-ethylenedioxythiophene)-Bi_2S_3 nanocomposites. *J. Nanoparticle Research.*, **14**, 812 (2012).
15. Y. Du, K. F. F. Cai, S. Z. Shen, B. J. An, Z. Qin, and P. S. Casey. Influence of sintering temperature on thermoelectric properties of BiTe/polythiophene composite materials. *J. Mater. Sci.: Mater. Elec.*, **23**, 870 (2012).
16. Y. Du, S. Z. Shen, K. Cai, and P. S. Casey, Research progress on polymer-inorganic thermoelectronic nanocomposite materials. *Progr. Polymer Sci.*, **37**, 820 (2012).
17. N. Toshima, N. Jiravanichanun, and H. Marutani, Organic thermoelectric materials composed of conducting polymer and metal nanoparticles. *J. Elect. Mater.*, **41**, 1735 (2012).
18. M. He, J. Ge, Z. Lin, X. Feng, X. Wang, H. Lu, Y. Yang, and F. Qiu. Thermopower enhancement in conducting polymer nanocomposites

via carrier energy scattering at the organic–inorganic semiconductor interface. *Energy Environ. Sci.*, **5**, 8351 (2012).

19. S. K. Yee, N. E. Coates, A. Majumdar, J. J. Urban, and R. A. Segalman, Thermoelectric powerfactor optimization in PEDOT:PSS tellurium nanowire hybrid composites. *Phys. Chem. Chem. Phys.*, **15**, 4024 (2013).

20. N. E. Coates, S. K. Yee, B. McCulloch, K. C. See, A. Majumdar, R. A. Segalman, and J. J. Urban, Effect of interfacial properties on polymer–nanocrystal thermoelectronc transport. *Adv. Mater.*, **25**, 1629 (2013).

21. M. Zebarjadi, K. Esfarjani, M. S. Dresselhaus, Z. F. Ren, and G. Chen. Perspectives on thermoelectronics: From fundamentals to device applications. *Energy Environ. Sci.*, **5**, 5147 (2012).

22. T. O. Poehler, and H. E. Katz. Prospects for polymer-based thermo-electrics: State of the art and theoretical analysis. *Energy Environ. Sci.*, **5**, 8110 (2012).

23. O. Bubnova, M. Berggren, and X. Crispin, Tuning the thermoelectric properties of conducting polymers in an electrochemical transistor. *J. Am. Chem. Soc.*, **134**, 16456 (2012).

24. G. H. Kim, L. Shao, K. Zhang, and K. P. Pipe, Engineered doping of organic semiconductors for enhanced thermoelectric efficiency. *Nature Mater.*, **12**, 719 (2013).

25. J. C. Duda, P. E. Hopkins, Y. Shen, and M. C. Gupta, Exceptionally low thermal conductivities of films of the fullerene derivative PCBM. *Phys. Rev. Lett.*, **110**, 015902 (2013).

26. O. Bubnova, and X. Crispin, Towards polymer-based organic thermo-electronic generators. *Energy Environ. Sci.*, **5**, 9345 (2012).

27. W. M. Qu, M. Plotner, and W. J. Fischer, Microfabrication of thermoelectric generators on flexible foil substrates as a power source for autonomous microsystems. *J. Micromech. Microeng.*, **11**, 146 (2011).

28. W. Glatz, E. Schwyter, L. Durrer, C. J. Hierold, Bi_2Te_3-based flexible micro thermoelectric generator with optimized design. *Microelectromech. Syst.*, **18**, 763 (2009).

29. K. Yazawa and A. Shakouri, Scalable cost/performance analysis for thermoelectric waste heat recovery systems. *J. Electron. Mater.*, **41**, 1845 (2012).

30. R. R. Sondergaard, M. Hosel, N. Espinosa, M. Jorgensen, and F. C. Krebs, Practical evaluation of organic polymer thermoelectrics by large-area R2R processing on flexible substrates. *Energy Sci. Eng.* **1**, 81(2013).

31. J. Wusten and K. Potje-Kamloth, Organic thermogenerators for energy autarkic systems on flexible substrates. *J. Phys. D: Appl. Phys.*, **41**, 135113 (2008).

32. T. Park, C. Park, B. Kim, H. Shin, and E. Kim, Flexible PEDOT electrodes with large thermoelectric power factors to generate electricity by the touch of fingertips. *Energy Environ. Sci.*, **6**, 788 (2013).

33. E. Hu, A. Kaynak, and Y. C. Li, Development of a cooing fabric from conducting polymer coated fibres: proof of concept. *Synth Met*, **150**, 139 (2005).

34. A. C. Sparavigna, L. Florio, J. Avloni, and A. Henn, Polypyrrole coated PET fabrics for thermal applications. *Mater. Sci. Appl.* **1**, 253 (2010).

35. E. J. X. Pang, A. Chan, and S. J. Pickering, Thermoelectrical properties of intercalated recycled carbon fibre composite. *Composites: Part A*, **42**, 1406 (2011).

36. O. Bubnova, Z. U. Khan, A. Malti, S. Braun, M. Fahlman, M. Berggren, and X. Crispin, Optimization of the thermoelectric figure of merit in the conducting polymer poly(3,4-ethylenedioxy-thiophene). *Nature Mater.*, **10**, 429 (2011).

37. L. Xu, Y. Liu, B. Chen, C. Zhao, and K. Lu, Enhancement in thermoelectric properties using a p-type and n-type thin-film device structure. *Polym. Compos.*, **34**, 1728 (2013).

38. Y. Sun, P. Sheng, C. Di, F. Jiao, W. Xu, D. Qiu, and D. Zhu, Organic thermoelectric materials and devices based on p- and n-type poly(metal thiolate)s. *Adv. Mater.*, **24**, 932 (2012).

39. L. Yan, M. Shao, H. Wang, D. Dudis, A. Urbas, and B. Hu, High S effects from hybrid metal/polymer/metal thin-film devices. *Adv. Mater.*, **23**, 4120 (2011).

40. M. Stanford, H. Wang, I. Ivanov, and B. Hu, High S effects from conducting polymer: PEDOT:PSS based thin-film device with hybrid metal/polymer/metal architecture. *Appl. Phys. Lett.*, **101**, 173304 (2012).

41. L. Xu, Y. Liu, M. P. Garrett, B. Chen, and B. Hu, Enhancing S effects by using excited states in organic semiconducting polymer MEH-PPV on multilayer electrode/polymer/electrode thin-film structure. *J. Phys. Chem.*, **117**, 10264 (2013).

42. C. A. Hewitt, A. B. Kaiser, S. Roth, M. Craps, R. Czerw, and D. L. Carroll, Multilayered carbon nanotube/polymer composite based thermoelectric fabrics. *Nano Lett.*, **12**, 1307 (2012).

43. M. Leclerc and A. Najari, Organic thermoelectronics: Green energy from a blue polymer. *Nature Mater.*, **10**, 409 (2011).

44. M. He, F. Qiu, and Z. Q. Lin, Towards high performance polymer-based thermoelectric materials. *Energy Environ. Sci.*, **6**, 1352 (2013).

45. M. Chabinyc, Thermoelectric polymers: Behind organics' thermopower. *Nature Mater.*, **13**, 119 (2014).

46. O. Bubnova, Z. U. Khan, H. Wang, S. Braun, D. R. Evans, M. Fabretto, P. Hojati-Talemi, D. Dagnelund, J. B. Arlin, Y. H. Geerts, S. Desbief, D. W. Breiby, J. W. Andreasen, R. Lazzaroni, W. M. M. Chen, I. Zozoulenko, M. Fahlman, P. J. Murphy, M. Berggren, and X. Crispin, Semi-metallic polymers. *Nature Mater.*, **13**, 190 (2014).

47. M. Scholdt, H. Do, J. Lang, A. Gall, A. Colsmann, U. Lemmer, J. Koenig, M. Winkler, and H. Boettner, Organic semiconductors for thermoelectric applications. *J. Electron. Mater.*, **39**, 1589 (2010).

48. J. Sun, M. L. Yeh, B. J. Jung, B. Zhang, J. Feser, A. Majumdar, and H. E. Katz, Simultaneous increase in S coefficient and conductivity in a doped poly(alkylthiophene) blend with defined density of states. *Macromolecules*, **43**, 2897 (2010).

49. Y. Xuan, X. Liu, S. Desbief, P. Leclere, M. Fahlman, R. Lazzaroni, M. Berggren, J. Cornil, D. Emin, and X. Crispin, Thermoelectric properties of conducting polymers: The case of poly(3-hexylthiophene). *Phys. Rev. B*, **82**, 115454 (2010).

50. O. T. Ikkala, L. O. Pietila, L. Ahjopalo, H. Osterholm, and P. J. Passiniem, On the molecular recognition and associations between electrically conducting polyaniline and solvents. *J. Chem. Phys.*, **103**, 9855 (1995).

51. S. Bhadra, D. Khastgir, N. K. Singha, and J. H. Lee, Progress in preparation, processing and applications of polyaniline. *Prog. Polym. Sci.*, **34**, 783 (2009).

52. Y. N. Sun, Z. M. Wei, W. Xu, and D. B. Zhu, A three-in-one improvement in thermoelectric properties of polyaniline brought by nanostructures. *Synth Met*, **160**, 2371 (2010).

53. N. Toshima, Conductive polymers as a new type of thermoelectric material. *Macromol. Symp.*, **186**, 81 (2002).

54. H. Yan, and N. Toshima, Thermoelectric properties of alternatively layered films of polyaniline and (+/-)-10-camphorsulfonic acid-doped polyaniline. *Chem. Lett.*, **28**, 1217 (1999).

55. H. Yan, T. Ohta, and N. Toshima, Stretched polyaniline films doped by ((+/-))-10-camphorsulfonic acid: Anisotropy and improvement of thermoelectric properties. *Macromol. Mater. Eng.*, **286**, 139 (2001).

56. X. Gao, K. Uehara, D. D. Klug, S. Patchkovskii, and T. M. Tse, Theoretical studies on the thermopower of semiconductors and low-band-gap crystalline polymers. *Phys. Rev. B*, **72**, 125202 (2005).

57. D. Vanderbilt, Soft self-consistent pseudopotentials in a generalized eigenvalue formalism. *Phys. Rev. B*, **41**, 7892 (1990).

58. G. Kresse and J. Hafner, Ab-initio molecular-dynamics for open-shell transition-metals. *Phys. Rev. B*, **48**, 13115 (1993).

59. K. Hiraishi, A. Masuhara, H. Nakanishi, H. Oikawa, and Y. Shinohara, Evaluation of thermoelectric properties of polythiophene synthesized by electrolytic polymerization. *J. Appl. Phys.*, **48**, 071501 (2009).

60. Y. Shinohara, Y. Isoda, Y. Imai, K. Hiraishi, H. Oikawa, and H. Nakanishi, The effect of carrier conduction between main chains on thermoelectric properties of polythiophene. In: Kim I, (ed.), *IEEE Proceedings ICT 07: Twenty-Sixth International Conference on Thermoelectrics.* pp. 410–412 (New York, 2008).

61. R. R. Yue, S. Chen, B. Y. Lu, C. C. Liu, and J. K. Xu, Facile electrosynthesis and thermoelectric performance of electroactive free-standing polythieno[3,2-b]thiophene films. *J. Solid State Electrochem.*, **15**, 539 (2011).

62. Y. Shinohara, Y. Imai, Y. Isoda, K. Hiraishi, and H, Nakanishi, A new challenge of polymer thermoelectric materials as ecomaterials. In: Chandra T, Tsuzaki K, Militzer M, Ravindran C, (eds.), *THERMEC 2006*, Pts. 1–5. Trans Tech Publications Ltd.: Stafa-Zurich: pp. 2329–2332 (2007).

63. I. Levesque, X. Gao, D. D. Klug, J. S. Tse, C. I. Ratcliffe, and M. Leclerc, Highly soluble poly(2,7-carbazolenevinylene) for thermoelectrical applications: From theory to experiment. *React. Funct. Polym.*, **65**, 23 (2005).

64. B. Y. Lu, C. C. Liu, S. Lu, J. K. Xu, F. X. Jiang, Y. Z. li, and Z. Zhang, Thermoelectric perfromances of free-standing polythiophene and poly(3-methylthiophene) nanofilms. *Chin. Pys. Lett.*, **27**, 057201 (2010).

65. T. Inabe, H. Ogata, Y. Maruyama, Y. Achiba, S. Suzuki, K. Kikuchi, and I. Ikemoto, Electronic structure of alkali metal doped C_{60} derived from thermoelectric power measurements. *Phys. Rev. Lett.*, **69**, 3797 (1992).

66. Z. H. Wang, K. Ichimura, M. S. Dresselhaus, G. Dresselhaus, W. T. Lee, K. A. Wang, and P. C. Eklund, Electronic transport properties of $K_x C_{70}$ thin films. *Phys. Rev. B*, **48**, 10657 (1993).

67. M. Sumino, K. Harada, M. Ikeda, S. Tanaka, K. Miyazaki, and C. Adachi, Thermoelectric properties of n-type C_{60} thin films and their application in organic thermovoltaic devices. *Appl. Phys. Lett.*, **99**, 093308 (2011).

68. T. Menke, D. Ray, J. Meiss, K. Leo, and M. Riede, In-situ conductivity and S measurements of highly efficient n-dopants in fullerene C_{60}. *Appl. Phys. Lett.*, **100**, 093304 (2012).

69. S. Kola, J. H. Kim, R. M. Ireland, M. L. Yeh, K. Smith, W. M. Guo, and H. E. Katz, Pyromellitic diimide-ethynylene-based homopolymer

film as an n-channel organic field-effect transistor semiconductor. *ACS Macro Lett.*, **2**, 664 (2013).

70. R. A. Schiltz, F. G. Brunetti, A. M. Glaudell, P. L. Miller, M. A. Brady, C. J. Takacs, C. J. Hawker, and M. L. Chabinyc, Solubility-limited extrinsic n-type doping of a high electron mobility polymer for thermoelectric applications. *Adv. Mater.*, **26**, 2825 (2014).

71. B. Russ, M. J. Robb, F. G. Brunetti, P. L. Miller, E. E. Perry, S. N. Patel, V. Ho, W. B. Chang, J. J. Urban, M. L. Chabinyc, C. J. Hawker, and R. A. Segalman, Power factor enhancement in solution-processed organic n-type thermoelectrics through molecular design. *Adv. Mater.*, **26**(11), 3473–3477 (2014).

72. Q. M. Li, L. Cruz, and P. Phillips, Granular-rod model for electronic conductivity in polyanaline. *Pys. Rev. B*, **47**, 1840 (1993).

73. N. Basescu, Z.-X. Liu, D. Moses, A. J. Heeger, H. Naarmann, and N. Theophilou, In Electronic properties of conjugated polymers, H. Kuzmany, M. Mehring, and S. Roth (eds.), Springer: Berlin, p. 18 (1987).

74. A. B. Kaiser, Thermoelectric power and conductivity of heterogeneous conducting polymers. *Phys. Rev. B*, **40**, 2806 (1989).

75. K. Bender, E. Gogu, I. hennig, D. Schweitzer, and H. Muenstedt, Electrical conductivity and thermoelectric power of various polypyrroles. *Synth. Met.*, **18**, 85 (1987).

76. G. Slack, In *CRC Handbook of Thermoelectrics*, edited by D. M. Rowe, CRC Press: Boca Raton, FL, pp. 407 (1995).

77. N. Tessler, Y. Preezant, N. Rapapaport, and Y. Roichman, Charge transport in disordered organic materials and its relevance to thin-film devices: a tutorial review. *Adv. Mater.*, **21**, 2741 (2009).

78. L. H. Jumison, Understanding microstructure and charge transport in semicrytalline polythiophenes. Ph. D. Dissertation, Mat. Sci. Eng., Stanford University, 2011.

79. A. B. Kaiser, Systematic conductivity behavior in conducting polymers: Effects of heterogeneous disorder. *Adv. Mater.*, **13**, 927 (2001).

80. H. Yan, N. Sada, and N. Tochima, Thermal transporting properties of electrically conductive polyanaline films as organic thermoelectric materials. *J. Therm. Anal. Calorim.*, **69**, 881 (2002).

81. F. F. Kong, C. C. Liu, J. K. Xu, F. X. Jiang, B. Y. Lu, R. R. Yue, G. D. Liu, and J. M. Wang, Simultaneous enhancement of σ and S coefficient of poly(3,4ethylenedioxythiophene):poly(styrenesulfonate) films treated with urea. *Chin. Phys. Lett.*, **28**, 037201 (2011).

82. F. X. Jiang, J. K. Xu, B. Y. Lu, Y. Xie, R. J. Huang, and L. F. Li, Thermoelectric performance of poly(3,4-ethylenedioxythiophene): poly(styrenesulfonate). *Chin. Phys. Lett.*, **25**, 2202 (2008).

83. M. Scholdt, H. Do, J. Lang, A. Gall, A. Colsmann, U. Lemmer, J. D. Koenig, M. Winkler, and H. Boettner, Organic semiconductors for thermoelectric applications. *J. Electron. Mater.* **39**, 1589 (2010).

84. J. Y. Kim, J. H. Jung, D. E. Lee, and J. Joo, Enhancement of σ of poly(3,4-ethylenedioxythiophene)/poly(4-styrenesulfonate) by a change of solvents. *Synth. Met.*, **126**, 311 (2002).

85. K. C. Chang, M. S. Jeng, C. C. Yang, Y. W. Chou, S. K. Wu, M. A. Thomas, and Y. C. Peng, The thermoelectric performance of poly(3,4-ethylenedioxythiophene)/poly(4-styrenesulfonate) thin films. *J. Electron. Mater.*, **38**, 1182 (2009).

86. R. J. Kline, M. D. McGehee, E. N. Kadnikova, J. Liu, and J. M. J. Frechet, Controlling the field-effect mobility of regioregular polythiophene by changing the molecular weight. *Adv. Mater.*, **15**, 1519 (2003).

87. J. Zhao, J. W. Jiang, and N. Wei, Thermal conductivity dependence on chain length in amorphous polymers. *J. Appl. Phys.*, **113**, 184304 (2013).

88. T. J. Kistenma, T. E. Phillips, and D. O. Cowan, Crystal-structure of 1-1 radical cation-radical anion salt of 2,2'-bis-1,3-dithiole and 7,7,8,8-tetracyanoquinodimethane. *Acta. Crystallogr. Sect. B: Struct. Sci.*, **30**, 763 (1974).

89. J. F. Kwak, P. M. Chaikin, A. A. Russel, A. F. Garito, and A. J. Heeger, Anisotropic thermoelectric-power of TTF-TCNQ. *Solid State Commun.*, **16**, 729 (1975).

90. M. Sakai, M. Iizuka, M. Nakamura, and K. Kudo, Organic nano-transistor fabricated by co-evaporation method under alternating electric field. *Synth. Met.*, **153**, 293 (2005).

91. E. Tamayo, K. Hayashi, T. Shinano, Y. Miyazaki, and T. Kajitani, Rubbing effect on surface morphology and thermoelectric properties of TTFTCNQ thin films. *Appl. Surf. Sci.*, **256**, 4554 (2010).

92. D. K. Taggart, Y. A. Yang, S. C. Kung, T. M. McIntire, and R. M. Penner, Enhanced thermoelectric metrics in ultra-long electrodeposited PEDOT nanowires. *Nano Lett.*, **11**, 125 (2011).

93. H. Dong, S. Jiang, L. Jiang, Y. Liu, H. Li, W. Hu, E. Wang, S. Yan, Z. Wei, W. Xu, and X. Gong, Nanowire crystals of a rigid rod conjugated polymer. *J. Am. Chem. Soc.*, **131**, 17315 (2009).

94. Y. Zhang, Z. Zhang, Y. Zhao, Y. Fan, T. Tong, H. Zhang, and Y. Wang, Solution-processed microwires of phthalocyanine copper(II) derivative with excellent conductivity. *Langmuir*, **25**, 6045 (2009).

95. E. V. Canesi, A. Luzio, B. Saglio, A. Bianco, M. Caironi, and C. Bertarelli, N-type semiconducting polymer fibers. *ACS Macro Lett.*, **1**, 366 (2012).

96. W. B. Huang, J. B. Peng, L. Wang, J. Wang, and Y. Cao, Impedance spectroscopy investigation of electron transport in solar cells based on blend film of polymer and nanocrystals. *Appl. Phys. Lett.*, **92**, 103308 (2008).

97. J. J. Li, X. F. Tang, H. Li, Y. G. Yan, and Q. J. Zhang, Synthesis and thermoelectric properties of hydrochloric acid-doped polyanaline. *Synth. Met.*, **4**, 2445 (2010).

98. F. Yakuphanoglu, B. F. Senkal, and A. Sarac, Thermoelectric power and optical properties of organosoluble polyaniline organic semiconductor. *J. Electron. Mater.*, **37**, 930 (2008).

99. T. Menke, D. Ray, J. Meiss, K. Leo, and M. Riede, In-situ conductivity and S measurements of highly efficient n-dopants in fullerene C60. *Appl. Phys. Lett.*, **100**, 093304 (2012).

100. M. L. Tietze, F. Wolzl, T. Menke, A. Fischer, M Riede, K. Leo, and B. Lussem, Self-passivation of molecular n-type doping during air exposure using a highly efficient air-instable dopant. *Physica Status Solidi A*, **210**, 2188 (2013).

101. M. T. Khan, A. Kaur, S. K. Dhawan, and S. Chand, *In-situ* growth of cadmium telluride nanocrystals in oly(3-hexylthiophene) matrix for photovoltaic application. *J. Appl. Phys.*, **110**, 044509 (2011).

102. L.-D. Zhao, V. P. Dravd, and M. G. Kanatzidis, The panoscopic approach to high performance thermoelectronics. *Energy Environ. Sci.*, **7**, 251 (2014).

103. R. B. Aich, N. Blouin, A. Bouchard, and M. Leclerc, Electrical and thermoelectric properties of poly(2,7-carbazole) derivatives. *Chem. Mater.*, **21**, 751 (2009).

104. Q. Zhang, H. Wang, Q. Zhang, W. Liu, B. Yo, H. Wang, G. Ni, G. Chen, and Z. Ren, Effect of silicon and sodium on thermoelectric properties of thallium-doped lead telluride-based materials. *Nano Lett.*, **12**, 2324 (2012).

105. A. Shakouri and S. Li, Thermoelectric power factor for electrically conductive polymers, in *Proceedings of the 18th International Conference on Thermoelectrics (ICT '99)*, pp. 402–406 (September 1999).

106. L. Fang, K. Zhou, F. Wu, Q. L. Huang, X. F. Yang, and C. Y. Kong, Effect of doping concentration on the thermoelectric properties of nano Ga-doped ZnO films, in *Proceedings of the 3rd International Nanoelectronics Conference (INEC '10)*, pp. 1175–1176, (January 2010).

107. Y. Nogami, H. Kaneko, T. Ishiguro, A. Takahashi, J. Tsukamoto, and N. Hosoito, On the metallic states in highly conducting iodine-doped polyacetylene. *Solid State Commun.*, **76**, 583 (1990).

108. J. H Lee, D.-S. Leem, H.-J. Kim, and J.-J. Kim, Effectiveness of p-dopants in an organic hole transporting material. *Appl. Phys Lett.*, **94**, 123306 (2009).

109. J. H. Lee, H.-M. Kim, K.-B. Kim, R. Kabe, P. Anzenbacher, and J.-J. Kim, Homogeneous dispersion of organic p-dopants in an organic semiconductor as origin of high charge generation efficiency. *Appl. Phys. Lett.*, **98**, 173303 (2011).

110. Q. Yao, L. D. Chen, W. Q. Zhang, S. C. Liufu, and X. H. Chen, Enhanced thermoelectric performance of single-walled carbon nanotubes/polyaniline hybrid nanocomposites. *ACS Nano*, **4**, 2445 (2010).

111. C. Z. Meng, C. H. Liu, and S. S. Fan, A promising approach to enhanced thermoelectric properties using carbon nanotube networks. *Adv. Mater.*, **22**, 535 (2010).

112. C. Yu, Y. S. Kim, D. Kim, and J. C. Grunlan. Thermoelectric behavior of segregated-network polymer nanocomposites. *Nano Lett.*, **8**, 4428 (2008).

113. D. Wang, L. Wang, W. Wang, X. Bai, and J. Li, Thermoelectric properties of poythiophene/MWCNT composites prepared by ball-milling. Proceedings of SPIE, **8409**, 151(2013).

114. K. Zhang, Y. Zhang, and S. Wang. Enhancing thermoelectric properties of organic composites through hierarchal nanostructures. *Sci. Rep.*, **3**, 3448 (2013).

115. C. H. Cartwright. Wiedmann–Franzshe number, thermal conductivity and thermoelectric power of tellurium. *Annalen Der Physik.*, **18**, 656 (1933).

116. S. K. Ramasesha and A. K. Singh. Thermoelectric power of tellurium under pressure up to 8 GPa. *Philos. Mag. B — Phys. Cond. Matter Stat. Mech. Electron. Opt. Magn. Prop.*, **64**, 559 (1991).

117. D. Kim, Y. Kim, K. Choi, J. C. Grunlan, and C. Yu, Improved thermoelectric behavior of nanotube-filled polymer composites with poly(3,4-ethylenedioxythiophene) poly(styrenesulfonate). *ACS Nano*, **4**, 513 (2010).

118. H. Bark, J. S. Kim, and H. Kim, Effect of multiwalled carbon nanotubes on the thermoelectric properties of a bismuth telluride matrix, *Current Appl. Phys.*, **13**, S111 (2013).

119. Y. Choi, Y. Kim, and S.-G. Park, Effect of the carbon nanotube type on the thermoelectric properties of CNT/Nafion nanocomposites. *Org. Electron.*, **12**, 2120 (2011).

120. R. V. Gregory, W. C. Kimbrell, and H. H. Kuhn, Conductive textiles. *Synth. Met.*, **28**, 823 (1989).

121. G. E. Collins and L. J. Buckley, Conductive polymer-coated fabrics for chemical sensing. *Synth. Met.*, **78**, 93 (1996).

122. L. J. Buckley and M. Eashoo, Polypyrrole-coated fibers as microwave and millimeter obscurants. *Synth. Met.*, **78**, 1 (1996).

123. A. Kaynak, L. Wang, C. Hurren, and X. Wang, Characterization of conductive polypyrrole coated wool yarns. *Fibers Polym.*, **3**, 24 (2002).

124. H. Dong, S. Prasad, V. Nyame, and W. E. Jones, Sub-micrometer conducting polyanaline tubes prepared from polymer fiber templates. *Chem. Mater*, **16**, 371 (2004).

125. H. Dong, and W. E. Jones, Preparation of submicron polypyrrole/ poly(methyl methacrylate) coaxial fibers and conversion to polypyrrole and carbon tubes. *Langmuir*, **22**, 11384 (2006).

126. A. Yadav, K. P. Pipe, and M. Shtein, Fiber-based flexible thermoelectric power generator. *J. Power Sources*, **175**, 909 (2008).

127. D. Vatansever, E. Siores, R. L. Hadimani, and T. Shah, Smart woven fabrics in renewable energy generation, Advances in modern woven fabrics technology, S. Vassiliadis (ed.), InTech, (2011).

128. L. M. Goncalves, J. G. Rocha, C. Couto, P. Alpuim, G. Min, D. M. Rowe, and J. H. Correia, Fabrication of flexible thermoelectric micro-coolers using planar thin-film technologies. *J. Micromech. Microeng.*, **17**, S168 (2007).

129. D. Madan, A. Chen, P. K. Wright, and J. W. Evans, Dispenser printed composite thermoelectric thick films for thermoelectric generator applications. *J. Appl. Phys.*, **109**, 034904 (2011).

130. A. Chen, D. Madan, P. K. Wright, and J. W. Evans, Dispenser-printed planar thick-film thermoelectric energy generators. *J. Micromech. Microeng.*, **21**, 104006 (2011).

131. S. J. Kim, J. H. We, and B. J. Cho, A wearable thermoelectric generator fabricated on a glass fabric. *Energy Environ. Sci.*, **7**, 1959 (2014).

132. Z. Lu, M. Layani, X. Zhao, L. P. Tan, T. Sun, S. Fan, O. Yan, S. Magdassi, and H. H. Hng, Fabrication of flexible thermoelectric thin film devices by inkjet printing. *Small*, **10**, 3551 (2014).

133. M. Feinaeugle, C. L. Sones, E. Koukharenko, and R. W. Eason, Fabrication of a thermoelectric generator on a polymer-coated substrate via laser-induced forward transfer of chalcogenide thin films. *Smart Mater. Struct.*, **22**, 115023 (2013).

134. P. M. Allemand, K. C. Khemani, A. Koch, F. Wudl, K. Holczer, S. Donovan, G. Gruner, and J. D. Thompson, Orgnic molecular soft ferromagnetism in a fullerene C_{60}. *Science*, **253**, 301 (1991).

135. A. F. Hebard, M. J. Rosseinsky, R. C. Haddon, D. W. Murphy, S. H. Glarum, T. T. Palstra, A. P. Ramirez, and A. R. Kortan, Conducting films of C_{60} and C_{70} by alkali-metal doping. *Nature*, **350**, 600 (1991).

136. A. A. Zakhidov, H. Araki, K. Tada, K. Yakushi, and K. Yoshino, Granular superconductivity in a conducting polymer–fullerene–alkali metal composite, *Phys. Lett. A*, **205**, 317 (1995).

137. N. W. Gothard, T. M. Tritt, and J. E. Spowart, Figure of merit enhancement in bismuth telluride alloys via fullerene-assisted microstructural refinement. *J. Appl. Phys.*, **110**, 023706 (2011).

138. M. Popov, S. Buga, P. Vysikaylo, P. Stepanov, V. Skok, V. Medvedev, E. Tatyanin, V. Denisov, A. Kirichenko, V. Aksenenkov, and V. Blank, C_{60}-doping of nanostructured Bi-Sb–Te thermoelectrics. *Physica Status Solidi A*, **208**, 2783 (2011).

139. X. Shi, L. Chen, J. Yang, and G. P. Meisner, Enhanced thermoelectric figure of merit of CoSb3 via large-defect scattering. *Appl. Phys. Lett.*, **84**, 2301 (2004).

140. F. Li, M. Pfeiffer, A. Werner, K. Harada, K. Leo, N. Hayashi, K. Seki, X. Liu, and X.-D. Dang, Acridine orange base as a dopant for n doping of C_{60} films. *J. Appl. Phys.*, **100**, 023716 (2006).

141. P. Wei, T. Menke, B. D. Naab, K. Leo, M. Riede, and Z. Bao, 2-(2-Methoxyphenyl)-1,3-dimethyl-1H-benzoimidazol-3-ium iodide as a new air-stable n-type dopant for vacuum-processed organic semiconductor thin films. *J. Amer. Chem. Soc.*, **134**, 3999 (2012).

142. A. A. Zakhidov, Weak charge transfer dopants in conducting polymers: Possible sensibilization of photoconductivity. *Synth. Met.*, **41**, 3393 (1991).

143. Y. Wang, Photoconductivity of fullerene-doped polymers. *Nature*, **356**, 585 (1992).

144. G. Yu, K. Pakbaz, and A. J. Heeger, Semiconducting polymer diodes: Large size, low cost photodetectors with excellent visible-ultraviolet sensitivity. *Appl. Phys. Lett.*, **64**, 3422 (1994).

145. W. U. Huynh, J. J. Dittmer, and A. P. Alivisatos, Hybrid nanorod-polymer solar cells. *Science*, **295**, 2425 (2002).

146. N. C. Greenham, X. G. Peng, and A. P. Alivisatos, Charge separation and transport in conjugated-polymer/semiconductor–nanocrystal composites studied by photoluminescence quenching and photoconductivity. *Phys. Rev. B: Condens. Matter Mater. Phys.*, **54**, 17628 (1996).

147. W. B. Huang, J. B. Peng, L. Wang, J. Wang, and Y. Cao, Impedance spectroscopy investigation of electron transport in solar cells based on blend film of polymer and nanocrystals. *Appl. Phys. Lett.*, **92**, 013308 (2008).

148. J. Seo, W. J. Kim, S. J. Kim, K. S. Lee, A. N. Cartwright, and P. N. Prasad, Polymer nanocomposite photovoltaics utilizing CdSe nanocrystals capped with thermally cleavable solubilizing ligand. *Appl. Phys. Lett.*, **94**, 133302 (2009).

149. Y. F. Zhou, F. S. Riehle, Y. Yuan, H. F. Schleiermacher, M. Niggemann, G. A. Urban, and M. Kruger, Improved efficiency of hybrid solar cells based on non-ligand-exchanged CdSe quantum dots and poly(3-hexylthiophene). *Appl. Phys. Lett.* **96**, 013304 (2010).

150. J. S. Liu, T. Tanaka, K. Sivula, A. P. Alivisatos, and J. M. Frechet, Employing end-functionalized polythiophene to control the morphology of nanocrystal–polymer composites in hybrid solar cells. *J. Am. Chem. Soc.*, **126**, 6550 (2004).

151. W. U. Huynh, X. G. Peng, and A. P. Alivisatos, CdSe nanocrystal rods/poly(3-hexylthiophene) composite photovoltaic device. *Adv. Funct. Mater.*, **11**, 923 (1999).

152. D. Yu, B. L. Wehrenberg, P. Jha, J. Ma, and P. Guyot-Sionnest, Electronic transport of n-type CdSe quantum dot films: Effect of film treatment. *J. Appl. Phys.*, **99**, 114315 (2006).

153. Y. Q. Li, R. Mastria, K. C. Li, A. Fiore, Y. Wang, R. Cingolani, L. Manna, and G. Gigle, Improved photovoltaic performance of bilayer heterojunction photovoltaic cells by triplet materials and tetrapod-shaped colloidal nanocrystal doping. *Appl. Phys. Lett.*, **95**, 043101 (2009).

154. S. Dayal, N. Kopidakis, D. C. Olson, D. S. Ginley, and G. Rumbles, Photovoltaic devices with a low band gap polymer and CdSe nanostructures exceeding 3% efficiency. *Nano Lett.*, **10**, 239 (2010).

155. I. Gur, N. A. Fromer, C. P. Chen, A. G. Kanaras, and A. P. Alivisatos, Hybrid solar cells with prescribed nanoscale morphologies based on hyperbranched semiconductor nanocrystals. *Nano Lett.*, **7**, 409 (2007).

156. S. Dayal, M. O. Reese, A. J. Ferguson, D. S. Ginley, G. Rumbles, and N. Kopidakis, The effect of nanoparticle shape on the photocarrier dynamics and photovoltaic device performance of poly(3-hexylthiophene): CdSe nanoparticle bulk heterojunction solar cells. *Adv. Funct. Mater.*, **20**, 2629 (2010).

157. E. Holder. N. Tessler, and A. L. Rogach, Hybrid nanocomposite materials with organic and inorganic components for opto-electronic devices. *J. Mater. Chem.*, **18**, 1064 (2008).

158. S. Gupta, Q. L. Zhang, T. Emrick, and T. P. Russell, "Self-corralling" nanorods under an applied electric field. *Nano Lett.*, **6**, 2066 (2006).

159. C. B. Murray, C. R. Kagan, and B. G. Bawendi, Synthesis and characterization of monodisperse nanocrystals and close-packed nanocrystal assemblies. *Annu. Rev. Mater. Sci.*, **30**, 545, (2000).

160. A. L. Rogach, D. V. Talapin, E. V. Shevchenko, A, Kornowski, M. Haase, and H. Weller, Organization of matter on different size scales: Monodisperse nanocrystals and their superstructures. *Adv. Funt. Mater.*, **12**, 653 (2002).

161. M. Brinkmann, D. Aldakov, and F. Chandezon, Fabrication of oriented and periodic hybrid nanostructures of regrioregular poly(3-hexylthiophene) and CdSe nanocrystals by directional epitaxial solidification. *Adv. Mater.*, **19**, 3819 (2007).

162. M. Brinkmann and J. C. Wittmann, Orientation of regioregular poly(3-hexylthiophene) by directional solidification: A simple method to reveal the semicrystalline structure of a crystalline polymer. *Adv. Mater.*, **18**, 860 (2006).

163. Z. Wu, A. Petzold, T. Henze, T. Thurn-Albrecht, R. H. Lohwasser, M. Sommer, and M. Thelakkat, Temperature and molecular weight dependent hierarchical equilibrium structures in semiconducting poly(3-hexylthophene). *Macromolecules*, **43**, 4646 (2010).

164. J. J. Xu, J. C. Hu, X. F. Liu, X. H Qiu, and Z. X. Wei, Stepwise self-assembly of P3HT/CdSe hybrid nanowires with enhanced photoconductivity. *Macromol. Rapid Commun.*, **30**, 1419 (2009).

165. H. Sirringhaus, P. J. Brown, R. H. Friend, M. M. Nielsen, K. Bechgaard, B. M. W. Langeveld-Voss, A. J. H. Spiering, R. A. J. Janssen, and E. W. Meijer, Microstructure-mobility correlation in self-organized, conjugated polymer field-effect transistors. *Synth. Met.*, **111**, 129 (2000).

166. G. Zotti, B. Vercelli, A. Berlin, M. Pasini, T. L. Nelson, R. D. McCullough, and T. Virgili, Self-assembled structures of semiconductor nanocrystals and polymers for photovoltaics. 2. Multilayers of CdSe nanocrystals and oligo(poly)thiophene-based molecules. Optical, Electrochemical, photoelectrochemical, and photoconductive properties. *Chem. Mater.*, **22**, 1521 (2010).

167. P. K. Sudeep, K. T. Early, K. D. McCarthy, M. Y. Odoi, M. D. Barnes, and T. Emrick, Monodisperse oligo(phenylene vinylene) ligands on CdSe quantum dots: Synthesis and polarization anisotropy measurements. *J. Am. Chem. Soc.*, **130**, 2384 (2008).

168. C. Fang, X. Y. Qi, Q. L. Fan, L. H. Wang, and W. Huang, A facile route to semiconductor nanocrystal-semiconducting polymer complex using amine-functionalized rod-coil triblock copolymer as multidentate ligand. *Nanotechnology*, **18**, 035704 (2007).

169. R. Pokrop, K. Pamula, S. Deja-Drogomirecka, M. Zagorska, J. Borysiuk, P. Reiss, and A. Pron, Electronic, electrochemical, and spectroelectrochemical properties of hybrid materials consisting of carboxylic acid derivatives of oligothiophene and CdSe semiconductor nanocrystals. *J. Phys. Chem. C*, **113**, 3487 (2009).

170. C. Querner, P. Reiss, M. Zagorska, O. Renault, R. Payerne, F. Genoud, P. Rannou, and A. Pron, Grafting of oligoanaline on CdSe nanocrystals: spectroscopic, electrochemical, and spectroelectro

chemical properties of the resulting organic/inorganic hybrid. *J. Mater. Chem.*, **15**, 554 (2005).

171. Q. Zhang, T. P. Russell, and T. Emrick, Synthesis and characterization of CdSe nanorods functionalized with regioregular poly(3-hexylthiophene). *Chem. Mater.*, **19**, 3712 (2007).

172. R. C. Shallcross, G. D. D'Ambruoso, J. Pyun, and N. R. Armstrong, Photoelectrochemical processes in polymer-tethered CdSe nanocrystals. *J. Am. Chem. Soc.*, **132**, 2622 (2010).

173. J. Xu, J. Wang, M. Mitchell, P. Mukherjee, M. Jeffries-El, J. W. Petrich, and Z. Q. Lin, Organic–inorganic nanocomposites via directly grafting conjugated polymers onto quantum dots. *J. Am. Chem. Soc.*, **129**, 12828 (2007).

174. M. D. Goodman, J. Xu, J. Wang, and Z. Q. Lin, Semiconductor conjugated polymer-quantum dot nanocomposites at the air/water interface and their photovoltaic performance. *Chem. Mater.*, **21**, 934 (2009).

175. R. Shenhar and V. M. Rotello, Nanoparticles: scaffolds and building blocks. *Acc. Chem. Res.*, **36**, 549 (2003).

176. C. R. van den Brom, I. Arfaoui, T. Cren, B. Hessen, T. T. M. Palstra, J. T. M. De Hosson, and P. Rudolf, Selective immobilization of nanoparticles on surfaces by loecular recognition using simple h-bonding functionalities. *Adv. Funct. Mater.*, **17**, 2045 (2007).

177. J. De Girolamo, P. Reiss, and A. Pron, Hybrid materials from diaminopyriminide-functionalized poly(hexylthiophene) and thymine-capped CdSe nanocrystals: Part II — Hydrogen bond assisted layer-by-layer molecular level processing. *J. Phys. Chem. C*, **112**, 8797 (2008).

178. A. A. Lutich, A. Poschl, G. X. Jiang, F. D. Stefani, A. S. Susha, A. L. Rogach, and J. Feldmann, Efficient energy transfer in layered hybrid organic/inorganic nanocomposites: A dual function of semiconductor nanocrystals. *Appl. Phys. Lett.*, **96**, 083109 (2010).

179. C. H. Chou, H. S. Wang, K. H. Wei, and J. Y. Huang, Thiophenol-modified CdS nanoparticles enhance the luminescence of benzoxyl Dendron-substituted polyfluorene copolymers. *Adv. Funct. Mater.*, **16**, 909 (2006).

180. P. A. van Hal, M. M. Wienk, J. M. Kroon, W. J. H. Verhees, L. H. Slooff, W. J. H. van Gennip, P. Jonkheijm, and R. A. J. Janssen, Photoinduced electron transfer and photovoltaic response of a MDMO-PPV: TiO_2 bulk-heterojunction. *Adv. Mater.*, **15**, 118 (2003).

181. W. J. E. Beek, M. M. Wienk, and R. A. J. Janssen, Hybrid solar cells from regioregular polythiophene and ZnO nanoparticles. *Adv. Funct. Mater.*, **16**, 1112 (2006).

182. S. D. Oosterhout, M. M. Wienk, S. S. van Bavel, R. Thiedmann, L. J. A. Koster, J. Gilot, J. Loos, V. Schmidt, and R. A. J. Janssen, The effect of three-dimensional morphology on the efficiency of hybrid polymer solar cells. *Nature Mater.*, **8**, 818 (2009).

183. A. Stavrinadis, S. Xu, J. H. Warner, J. L. Hutchison, J. M. Smith, and A. A. R. Watt, Superstructures of PbS nanocrystals in a conjugated polymer and the aligning role of oxidation. *Nanotechnology*, **20**, 445608 (2009).

184. H. C. Leventis, S. P. King, A. Sudlow, M. S. Hill, K. C. Molloy, and S. A. Haque, Nanostructured hybrid polymer-inorganic solar cell active layers formed by controllable *in situ* growth of semiconducting sulfide networks. *Nano Lett.*, **10**, 1253 (2010).

185. X. M. Peng, L. Zhang, Y. W. Chen, F. Li, and W. H. Zhou, In situ preparation and fluorescence quenching properties of poythiophen/ZnO nanocrystals hybrids through atom-transfer radical polymerization and hydrolysis. *Appl. Surf. Sci.*, **256**, 2948 (2010).

186. D. S. Ginger and N. C. Greenham, Charge injection and transport in films of CdSe nanocrystals. *J. Appl. Phys.*, **87**, 1361 (2000).

187. Y. Liu, M. Gibbs, J. Puthussery, S. Gaik, R. Ihly, H. W. Hillhouse, and M. Law, Dependence of carrier mobility on nanocrystal size and ligand length in PbSe nanocrystal solids. *Nano Lett.*, **10**, 1960 (2010).

Chapter 7

Modeling Thermoelectric Materials

Greg Walker

7.1 Introduction

The canonical formula for determining the efficacy of a materials system to perform as a thermoelectric energy conversion material is given by the dimensionless figure of merit,

$$ZT = \frac{S^2\sigma}{k_e + k_p}T. \tag{7.1}$$

This expression is useful because it (7.1) is a simple combination of transport properties and (7.2) provides a robust way to compare the performance of different materials systems. Despite its simplicity, the expression applies equally well to homogeneous bulk crystals (pure substances and alloys), nanostructured materials (superlattices and nanocrystalline composites), complex molecules (skutterudites), or any other structured materials system. Therefore, the application of this expression is considered universal, which is why we can compare different materials systems in a robust way. The origin of this universality is easy to understand. The dimensionless group can be derived from a global energy balance of a thermoelectric device. In other words, the figure of merit is based on a fundamental law of nature. This means that not only is the expression generally applicable, but also we can deduce physical understanding of why a material might make a good thermoelectric material by examining each factor independently.

Yet this innocuous collection of material properties belies the complexity in determining, through modeling, the transport properties inherent to thermoelectric materials. For example, in expressing the figure of merit in Eq. (7.1), we have employed a version that considers the thermal conductivity due to electron transport (k_e) and phonon transport (k_p) separately. For most bulk materials, thermal conductivity can be determined by looking up measured or accepted values from a suitable reference. If the thermal conductivity has been measured, the value represents the contribution from both electrons and phonons because measurements cannot distinguish the different components. Consequently, this separation of the components appears to be pedantic and unnecessary. If we wish *to model* the performance, however, both electronic and thermal systems must be considered. The transport in semiconductors, which make some of the best thermoelectric materials, will not be dominated by electrons or phonons; both contribute to transport. In fact, this characteristic is what makes them such good thermoelectric candidates. So we must solve for two energy-carrying systems. And these two systems are usually coupled, so that they must be solved simultaneously. Because electrons and phonons differ in mean free path lengths, models to describe the behavior inherently require multiscale methods to solve. So we see that very quickly a simple material property evaluation converts to a multiscale coupled simulation when we try to estimate the performance from physical models. This section is devoted to describing some of the prevailing methods for estimating the transport properties in modern thermoelectric materials systems. We will consider both electronic and thermal systems, but will focus the discussion primarily on noncontinuum types of models including particle-based methods (Boltzmann), quantum approaches, and atomistic simulations. Although the energy carriers — electrons and phonons — must usually be solved simultaneously, we will consider each system independently treating the coupling superficially in most cases. Assume that a very large and active community is working on the coupling problem, so we should not downplay the importance or difficulty of that part of the problem.

7.2 Continuum Considerations

The parameters and typical units that comprise Eq. (7.1) include the Seebeck coefficient (S, $\mu V K^{-1}$), the electrical conductivity (σ, $\Omega \, cm$), and thermal conductivity (k, $W m^{-1} K^{-1}$). The characteristic temperature T is usually the cold-side junction temperature and is used to nondimensionalize the quantity Z. Its exact value is unimportant for the foregoing discussion except to note that it represents some device reference temperature in the problem. The important thing to note is that each of these quantities represents a proportionality between a transport property and a driving potential. So, if we can calculate the transport due to the driving potential, then we can estimate the thermoelectric performance of a material.

7.2.1 Electrical conductivity

The electrical conductivity (reciprocal of the resistivity) is the proportionality of electrical current for a given electrical bias or field.

$$J = \sigma E. \tag{7.2}$$

The quantity can readily be deduced from IV characteristics like those shown in Fig. 7.1. The slope of any of the lines where the current is zero is proportional to the conductivity. Since the current density is given in the figure, the slope (dJ/dV) must be scaled by the area and length of the device so that

$$\sigma = \frac{A \, dJ}{l \, dV}, \tag{7.3}$$

which is essentially Ohm's law. In Fig. 7.1, the $\Delta T = 0$ line (dark blue) is linear close to the origin. Therefore, the slope can be approximated as a difference. At $V = 0.01 \, V$, the current is nearly

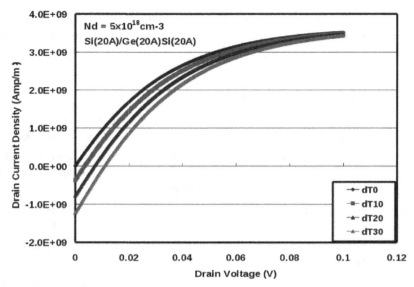

Figure 7.1. *IV* characteristics for a thermoelectric device. (Previously presented in Ref. [2].)

$J \approx 1 \times 10^9 \, \mathrm{A\,m^{-2}}$. For this $l = 6\,\mathrm{nm}$ cube of silicon material, the conductivity that results is $\sigma = 600 \, (\Omega\,\mathrm{m})^{-1}$. The actual device and model needed to generate the *IV* curves will be discussed later.

But are complex transport models really necessary? After all, using kinetic arguments, the conductivity of a material can be expressed as a simple collection of seemingly fundamental quantities.[1]

$$\sigma = \frac{ne\tau}{m}, \tag{7.4}$$

where n is the number of electrons, e is the electron charge, and m is the electron mass. However, this expression is known to be orders of magnitude off especially for nanostructured materials. In addition, the relaxation time τ is essentially not measurable. It depends on so many other physical variables that a universal expression is not available. In fact, transport models are required to obtain decent estimates of the relaxation time, so we usually have limited alternatives to calculating real transport.

7.2.2 Seebeck coefficient

The Seebeck coefficient is the proportionality between electrical bias created from a temperature difference. Admittedly, this quantity is not transport strictly speaking, but a value is easily derived from transport models. The Seebeck coefficient is obtained from IV characteristics (Fig. 7.1) by finding the electrical bias required to balance the temperature-induced voltage. That means we read the voltage at the zero current line for a given temperature difference across the device. In Fig. 7.1, the IV characteristic crosses the zero current line at $V \approx 0.01\,$V for an applied temperature of $\Delta T = 30\,$K. Therefore, the Seebeck coefficient is calculated to be $S = \Delta V/\Delta T = 333\,\mu\,V\,K^{-1}$ (note the unit conversion).

The combined parameter of the Seebeck coefficient squared and electrical conductivity comprise the power factor ($PF = S^2\sigma$). Although many past researchers have looked for good thermoelectrics by identifying materials with large Seebeck coefficients only, more recent efforts have realized the utility of identifying materials with large power factors. For our example device, $PF = 6.65 \times 10^{-5}\,A\,K^{-1}\,m^{-1}$.

7.2.3 Thermal conductivity

The thermal analog to the IV characteristics would be a $q\Delta T$ characteristics. From a plot of the heat flux (q) as a function of the driving temperature (ΔT), we can obtain the thermal conductivity using Fourier's law.

$$q = -kA\frac{dT}{dx}, \tag{7.5}$$

where k is simply a proportionality constant. Again, in a bulk sense, we have a chicken and egg problem; we need the thermal conductivity to obtain the heat flux. In a noncontinuum sense, however, we can obtain transport using more fundamental quantities, which will be described subsequently.

For both systems (electrical and thermal), there is no universal set of fundamental quantities used to predict either conductivity. Despite what we teach undergraduate engineering students, thermal and electrical conductivity are, in general, *not* material properties. In other words, we cannot always estimate the property by knowing the constituents of the materials systems only. We need the arrangement of the materials in the system, size and shape of the system, defect type and distribution, and chemical bonding, among a myriad of other physics that help fully describe the transport. The problem is that both electrical and thermal conductivity are effective properties derived from phenomenological models. Nevertheless, the subsequent sections demonstrate how noncontinuum models can be used to extract the requisite transport information.

7.3 Noncontinuum Methods

7.3.1 Particle-based transport

The celebrated Boltzmann transport equation (BTE) provides us with a framework for estimating transport from a particle point of view. For thermal transport, the particle is called a phonon; in electrical transport, we consider electrons. Although most readers are familiar with electrons, the phonon usually requires additional explanation. A phonon is a quantum particle in the same family with photons, and we conceptualize transport in a system as a result of a collection of heat-carrying particles moving in a solid. Assume that a phonon is *not* an atom in a lattice, even though the atom is a convenient particle to consider. Instead, it is more appropriate to think of a phonon as a vibrational mode of a collection of atoms in a periodic lattice. That mode, then, transports energy in a phonon packet much the way light is said to travel in wave packets. This picture of the phonon is important to grasp because there are many similarities between phonons and photons, which means we can leverage the vast literature on photon transport in participating media to devise solutions to phonons.

In its purest form, the BTE, which governs phonon or electron transport, is a first-order, seven-dimensional — three-space ($\mathbf{r}$), three-momenta ($\mathbf{p}$), and one-time (t) coordinates — nonlinear, integrodifferential equation, which does not lend itself to analytic or numerical solutions.

$$\frac{\partial f}{\partial t} \mathbf{b}\dot{\nabla}_r f + \mathbf{F}\dot{\nabla}_p f = G + \sum_{\mathbf{p}'} f(\mathbf{p}')[1 - f(\mathbf{p})]S(\mathbf{p}', \mathbf{p})$$

$$- \sum_{\mathbf{p}'} f(\mathbf{p})[1 - f(\mathbf{p}')]S(\mathbf{p}, \mathbf{p}'), \qquad (7.6)$$

where G is a source term and the summations describe the scattering into and out of different states. Moreover, the dependent variable f is a distribution function, which has little physical meaning, although it can be loosely interpreted as the probability of finding a particle at a specified location with specified momentum at a particular time. As such, researchers have simplified the BTE usually by taking moments and by invoking the relaxation time approximation (RTA).

Moments of the BTE reduce the dimensionality by 3 by removing the momentum or frequency space as an independent variable. The resulting quantities are averages over momentum. This approach can be quite good for systems where the characteristic lengths are large so that local thermodynamic equilibrium can be guaranteed. Usually device dimensions are compared to the mean free path of the particles to determine how small a system can be made before the local thermodynamic equilibrium breaks down. The Knudsen number $(Kn = \lambda/l)$, which is the ratio of the particle mean free path to a characteristic length of the device, provides a gross indication of whether bulk approximations are appropriate (for $Kn < 0.1$), or whether noncontinuum effects may dominate $(Kn > 0.1)$. The inequalities provide a gross rule of thumb; they are not strict rules. In practice, bulk relations often provide sufficient approximation to transport for Knudsen numbers up to unity, but the threshold depends on other characteristic lengths inherent to the system and how much accuracy is needed in the calculation.

The second approximation, the RTA, is invoked to cast the scattering integral (the summations) into a tractable form. The scattering integral includes the probabilistic interactions between all particles at all frequencies and modes. To compute a term exactly, the population of all states and the momentum- and energy-dependent scattering rates between any two states must be known. In general, this information is not available, although crude analytic approximations can be obtained from Fermi's golden rule. Scattering serves to return a nonequilibrium system back to equilibrium. Consequently, the formulation of the RTA merely compares the actual distribution to an equilibrium distribution scaled by a relaxation time.

$$\frac{f_0 - f}{\tau}. \tag{7.7}$$

This approximation is strictly valid only near equilibrium but in practice works remarkably well for a wide range of problems and dimensions. The reason why this approximation works so well is that the scattering processes are remarkably complex and extremely difficult to capture in their entirety. Therefore, the RTA, which combines all the frequency-dependent information into a single parameter, represents the integral of all the scattering physics involved. The single parameter, namely the average relaxation time for phonons or τ, is an effective quantity and is often considered a fitting parameter. So the relaxation time is an effective property, much the same way the thermal conductivity from Fourier's law is an effective property.

The relaxation time should not be interpreted as a mean free time — the average time between scattering events. Instead, several scattering events may have to occur before the nonequilibrium distribution is changed significantly. The relaxation time then represents the time required to push the nonequilibrium distribution back to the equilibrium configuration, which could be tens of mean free scattering times. Moreover, the value chosen for the relaxation time depends on whether we are interested in the relaxation of energy or momentum, which can be different.

With the RTA, the BTE can be recast into a form that effectively describes how an ensemble of particles, given by the distribution function f, evolves under various influences.

$$\frac{\partial f}{\partial t} + \mathbf{v} \cdot \nabla_{\mathbf{r}} f + \mathbf{F} \cdot_{\mathbf{p}} f = \frac{f_0 - f}{\tau(\mathbf{p}, \mathbf{r})} + G. \tag{7.8}$$

The first term is a transient term that provides for accumulation or loss of probability in time. The second term describes how probability is advected, much the way energy or mass is transported by the bulk motion of a fluid. The third term provides for the effects of external forces. The two terms on the right-hand side provide for interaction between particles within the distribution (RTA term) and interaction between particles outside the distribution (source term) such as in electron–phonon scattering.

If we assume that our system is close to equilibrium, we can make some modest simplifying assumptions that yield an expression for the distribution function. In particular, we neglect transient terms and assume that the gradient of the distribution function is dominated by the gradient in the equilibrium distribution. The linearized solution can now be written as

$$f = f_0 - \tau(\vec{v} \cdot \nabla_r f_0 + \vec{F} \cdot \nabla_p f_0). \tag{7.9}$$

From this expression we can derive relationships for the three material parameters: Seebeck coefficient, electrical conductivity, and thermal conductivity.

To obtain the Seebeck coefficient, we must first find the current density by taking the first moment of Eq. (7.9). This means we multiply the distribution function by momentum to the first power and sum over all momentum states,

$$J = \frac{1}{V} \sum_{k_x=-\infty}^{\infty} \sum_{k_y=-\infty}^{\infty} \sum_{k_z=-\infty}^{\infty} -evf. \tag{7.10}$$

The volume (V) and electron charge (e) convert the transport of particles to the transport of charge per volume. We can interpret the expression as a counting of the number of electrons traveling through the system as they also move in all directions. The current then will be the net motion of charge. We can convert the sum to an integral and integrate over the solid angle after plugging in Eq. (7.9) to obtain

$$J = -\frac{e}{3} \left(\frac{dE_f}{dx} + e\varepsilon \right) \int_0^{\infty} \tau v^2 D(E) \frac{\partial f_0}{\partial E} dE. \tag{7.11}$$

Here, $D(E)$ is the density of states and f_0 is the equilibrium distribution for electrons or the Fermi–Dirac distribution. The two components in parenthesis represent the two different forcing functions for electrons. The first term is the diffusion component because it is driven by spatial variations in the charge concentration. The second term is the drift term because the electrons are driven by the electric field E.

The Seebeck coefficient is essentially the diffusion of electrons due to temperature variations. To introduce the temperature into the diffusion component, we define the derivative of the distribution function as

$$\frac{\partial f_0}{\partial x} = -\frac{\partial f_0}{\partial E}\frac{dE_f}{dx} - \frac{E - E_f}{T}\frac{\partial f_0}{\partial E}\frac{dT}{dx},$$

$$J = -\frac{e}{3}\left(\frac{dE_f}{dx} + e\varepsilon\right)\int_0^\infty \tau v^2 D(E)\frac{\partial f_0}{\partial E}dE. \tag{7.12}$$

When this is applied to Eq. (7.11), we get three terms: two that contain the electrical components (electric field and charge concentration) and one that contains a thermal component. The Seebeck coefficient is defined as the ratio of the electrically driven electrons to the thermally driven electrons and is written as

$$S = -\frac{1}{eT}\frac{\int \tau v^2 (E - E_F)\left(\frac{\partial f_o}{\partial E}\right)D(E)dE}{\int \tau v^2 \left(\frac{\partial f_o}{\partial E}\right)D(E)dE},$$

$$J = -\frac{e}{3}\left(\frac{dE_f}{dx} + e\varepsilon\right)\int_0^\infty \tau v^2 D(E)\frac{\partial f_0}{\partial E}dE. \tag{7.13}$$

This expression is often called the Onsager relations, which are a collection of ratios of components of transport mechanisms.

7.3.2 Quantum transport

Electronic and thermal transport in highly scaled systems necessarily involves quantum transport. If materials structures approach nanoscale sizes such as in superlattices where the layer sizes are commensurate with a handful of atomic layers, then quantum mechanical confinement can affect the available states, the population of those states, transport within those states, and scattering between those states. At this point we are not considering what happens to the atomic scale, we only want to understand the effects of a highly scaled device. As such, we will employ an effective mass Hamiltonian where the physical parameters required are the effective mass and dielectric constants. These still are not fundamental physical quantities and require approximation themselves. Nevertheless, we can incorporate discretized energy levels, which is something that was essentially inaccessible with particle-based methods. This feature is important particularly for Seebeck calculations because the Seebeck effect can be adjusted through the filtering of different energy levels. In fact, one of the main reasons why nanoscale technology is being explored for thermoelectric materials is dramatic improvements in Seebeck coefficients through quantum confinement.

7.3.2.1 *Electrons*

The model for quantum electron transport is a Schrodinger/Poisson solver. Periodic solutions to these types of systems are prevalent because of the ease of implementation. By definition, however, periodic systems are not able to compute transport directly because we need some sort of driving potential to induce a bulk motion of the electrons, and a driving potential is impossible in a periodic system. To obtain transport properties in a periodic system, though, we can perform an autocorrelation of the electronic wavefunctions as a function of time. As a physical interpretation assume that the sloshing of electrons back and forth in a periodic system is related to the actual transport of electrons through a system. In other words, the local dynamics in each case are governed by the same physical processes. One primary drawback to this approach is that we cannot look at the effects of boundary conditions on transport, which is important in real devices. Therefore, we will consider a Green's function approach where real bias (both thermal and electrical) can be applied and transport quantities emerge naturally.

Another advantage of Green's function approach is its ability to include scattering in quantum systems. Only for the smallest most pure devices can scattering be neglected as an acceptable approximation. Normally, we do not think of a quantum system (wave-like transport) being able to incorporate scattering (particle behavior). However, if we understand at least on a conceptual level how Green's functions work, then we can appreciate how scattering can naturally be included in the framework.

Green's functions give the response to an impulse excitation. The impulse is a Dirac delta, which is trivially integrable. Therefore, Green's functions are often obtained analytically for homogeneous problems except where the impulse is the only nonhomogeneity. To obtain the full solution of a biased system, Green's function solution is integrated over the location and time of the nonhomogeneous boundary conditions, source conditions, or initial conditions. This approach is general enough to apply to our Schrödinger system as well as any other linear differential operator. But the point is that because Green's function represents an impulse instantaneous in time and localized in space, we can use this to represent a scattering event, which serves to couple the energy levels in the domain.

Although the formalism used here has been recognized for many years, the implementation for highly scaled devices was championed by a pair of electrical engineering professors at Purdue — Mark Lundstrom and Supriyo Datta. The approach described here follows their development with the exception of the application of temperature boundary conditions for the explicit calculation of thermoelectric properties.

An isolated device is modeled using Schrödinger's equation, where φ is the eigenstate of the electron.

$$(H + U)\psi_\alpha(r) = \varepsilon_\alpha \psi_\alpha(\mathbf{r}). \tag{7.14}$$

The Hamiltonian and the Hartree potential are H and U, respectively, and α indicates which eigenstate is being solved. Each eigenstate is solved independently in a ballistic (no scattering) simulation.

Once the eigenstates are found, the electron density matrix in real space is given by [3]

$$\rho(\vec{r}, \vec{r}') = \int_{-\infty}^{\infty} f_0(E - \mu)\delta(EI - H)dE, \tag{7.15}$$

where f_0 is the Fermi function, and $\delta(EI - H)$ identifies the energies associated with the device Hamiltonian, and here is where the impulse is introduced to formulate Green's function.

Using an expansion of the delta function, we obtain

$$\delta(EI - H) = \frac{i}{2\pi}\{[(E - i0^+)I - H]^{-1} - [(E + i0^+)I - H]^{-1}\},$$

$$= \frac{i}{2\pi}[G(E) - G^+(E)],$$

$$J = -\frac{e}{3}\left(\frac{dE_f}{dx} + e\varepsilon\right)\int_0^{\infty} \tau v^2 D(E)\frac{\partial f_0}{\partial E}dE, \tag{7.16}$$

such that the retarded and advanced Green's functions are defined as

$$G^{\mp}(E) = [(E \mp i0^+)I - H]^{-1}. \tag{7.17}$$

The impulse is essentially an electron density at a particular energy. In the energy domain, Green's function gives the energy eigenvalues for the eigenstates that are occupied in response to the applied impulse.

The spectral function, which is found from Green's functions, is useful to calculate several intermediate quantities,

$$A(r, r^t, E) = 2\pi i[G(E) - G^+(E)]. \tag{7.18}$$

For example, the diagonal elements of the spectral function represent the local electron density of states. In addition, the spectral function can be used to solve directly for transmission coefficients.

The density solution (Eq. (7.15)) is solved self-consistently with Poisson's equation,

$$\nabla^2 U = -\frac{q}{\varepsilon_s}(N_d - n),$$ (7.19)

where n is the real diagonal of the density matrix ρ, N_d is the donor level, and ε_s is the permittivity. This framework for an isolated device must be coupled to the contacts for meaningful engineering devices.

By adding self-energy terms to represent the bias added at the contacts, we can reformulate Eqs. (7.14) and (7.17) as

$$[H + U + \Sigma_1 + \Sigma_2]\psi_\alpha(\mathbf{r}) = \varepsilon_\alpha \psi_\alpha(\mathbf{r}),$$ (7.20)

$$G(E) = [(E - i0^+)I - H - \Sigma_1 - \Sigma_2]^{-1}.$$ (7.21)

We introduce a convenience term

$$\Gamma = i(\Sigma - \Sigma^+)$$ (7.22)

to represent the shifting of energy levels due to the contact (real component) and broadening of the channel levels due to the contact (imaginary component). Now the density is expressed in terms of these new terms as

$$\rho = \frac{1}{2\pi} \int_{-\infty}^{\infty} G[\Gamma_1 f_1 + \Gamma_2 f_2]G^+ dE,$$ (7.23)

where f is the Fermi function of the corresponding contacts.

The results of the solution give Green's function, which can be used to calculate the spectral function. The transport (or electrical current in this case) at a given contact is the sum of the flow of electrons from the channel to the contact and from the contact to the channel,

$$I_i = \frac{-q}{\hbar} \int_{\infty}^{\infty} \text{Trace}[\Gamma_i A]f_i - \text{Trace}[\Gamma_i G^n]dE.$$ (7.24)

Now we can see how to generate an *IV* curve; we simply change the applied voltage in the Fermi function and calculate the resulting current. The Fermi function also contains a temperature, so we can impose a temperature bias as easily as we add a voltage bias. From the *IV* curves, which can be found as a function of temperature, comes the thermoelectric properties as described in the discussion of bulk models. However, in this case, we have incorporated quantum effects found in nanoscale devices.

7.3.2.2 *Phonons*

Quantum transport of phonons[3] follows that of the transport of electrons using Green's functions with two conceptual differences. Phonons are bosons, which do not conserve number. Therefore, instead of the Fermi function, we use the Bose–Einstein distribution. Also, because phonons do not conserve number and do not have an inherent forcing function, we can dispense with Poisson's equation. When Green's function approach is used to calculate phonon systems, the technique has also been called the atomic Green's function formalism. Although the details of the calculation are similar to the electron model, a thorough treatment is left to other articles (e.g., Ref. [4]).

Without solving for the transport of phonons, however, we can still calculate their effect on electron transport through the self-scattering matrix. That is, we introduce a Σ_s into Eq. (7.21) alongside Σ_1 and Σ_2 for the contacts. We interpret the new matrix as a channel broadening of the levels that serves to reduce the transport in the channel. The matrix is populated with scattering rates, which can be energy dependent. For simplicity, we can use a constant scattering rate in the spirit of the relaxation time approximation to demonstrate the effects of scattering on the transport and ultimate thermoelectric property calculation.

Figure 7.2 is provided to illustrate the importance including scattering (presumably via phonons) in quantum systems. The system is a thin film of silicon and the current through the film is calculated at a fixed electric field. The analysis does not include phonon transport, but electron scattering is included through bulk scattering rates in the self-scattering matrix. Obviously for the smallest film thicknesses the scattering is negligible because the mean free path of phonons is

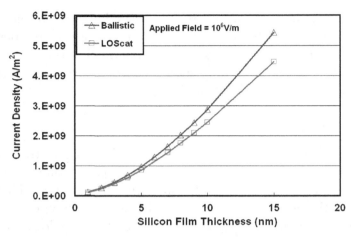

Figure 7.2. Current density in films of varying thickness. The film is doped at 10^{18} cm^{-3} for (a) ballistic transport and (b) electron transport with longitudinal optical-phonon scattering. Bulusu and Walker, *Superlattices and Microstructures*, **44**, 1–36, (2008).

quite a bit larger than the film. Therefore, transport is primarily ballistic at the smaller film sizes. As the film thickness increases, though, the influence of scattering even in quantum systems becomes apparent. In the ballistic case, the current rises because the quantum levels become closer to each other as well as decrease in energy. This allows more electrons to occupy the channel and contribute to transport. Theoretically, the ballistic current will increase without bound as the film thickness becomes larger. In reality, the scattering prevents an unbounded current, which will level off at larger film sizes. The point at which the current saturates, however, is well beyond that shown in Fig. 7.2. We would expect saturation at several times the mean free path, which is of the order of 100 nm for silicon. Figure 7.2 also illustrates another important concept for the nanostructured thermoelectric material design.

For highly scaled material structures, less than 5–7 nm, scattering does not play a role in the transport. We might assume that this would be beneficial for thermoelectric performance because good thermoelectrics need good electrical conductivity, and scattering limits the electrical conductivity. However, the level filtering introduced by the quantum mechanics prevents a large amount of current, so the capacity of these material structures is hampered by the quantum mechanical confinement. Therefore, smaller is not always better, and an optimized structure that balances the scattering against the confinement should outperform a structure that was blindly made smaller. This concept becomes even more important when considering phonon transport because the interfaces act as scattering sites and more interfaces tend to reduce the thermal conductivity.[6] So good design is a true balancing act between maximizing electron transport and minimizing phonon transport.

7.4 Wave versus Particle

The preceding text suggests that two independent methods (Boltzmann and quantum) can be used to calculate the transport in a nanostructured thermoelectric material. However, the two methods are indeed incompatible, and it is not obvious how the two disparate systems could both produce valid results for the motion, distribution, and energy of charge carriers. Assume that both are expressions of conservation laws, but the physics behind each is fundamentally different. Therefore, it is important to explore the space where each is applicable. In general, quantum transport or wave-like transport is preferable for systems where the characteristic dimensions are of the order of the wavelength of the carriers. Particle-based transport models are better when the characteristic dimensions are of the order of the mean free path of the carriers. Unfortunately, there is no universal rule about what model will dominate for a given application.

In Fig. 7.3, we show the electron transport as predicted by the NEGF ballistic model and a Boltzmann model in the relaxation time approximation with constant scattering rates (CRTA) in a 6 and a 12 nm silicon film.

Note that both models predict the same trend as a function of doping. With more carriers, we get more transport. However, the notable difference between the two calculation approaches is the disparity between the one- and two-band models. The number of bands was chosen based on the proximity of the levels to the conduction

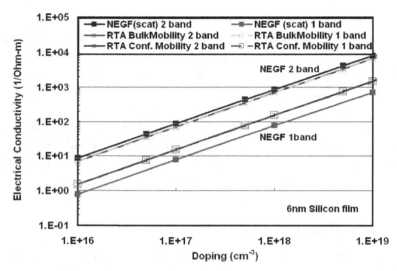

Figure 7.3. Electrical conductivity as calculated for a 6 nm film using both the NEGF and Boltzmann model in the relaxation time approximation (CRTA). Bulusu and Walker, *Superlattices and Microstructures*, **44**, 1–36, (2008).

band. The CRTA model shows very little transport in the second band, whereas the NEGF model predicts that both bands support nearly the same amount of current for the 6 nm film. Similar trends, although not as dramatic, are found in the 12 nm film. The problem is that confinement effects are largely obtained from the electron mobility,

$$\sigma = \frac{1}{2\pi a} \left(\frac{2k_B T}{\hbar^2} \right) \sqrt{m_x m_y} F_0 e\mu. \tag{7.25}$$

To obtain confined conductivity from (7.25), one must first obtain the confined mobility μ. This must be obtained from experiments where there is still debate over the appropriate value.[7–10] Consequently, the Boltzmann model can be made to estimate transport accurately, but may require additional information that is not always available and difficult to obtain with any reliability. Therefore, it may not be able to cope with advances being made in nanostructured materials for thermoelectric applications.[5]

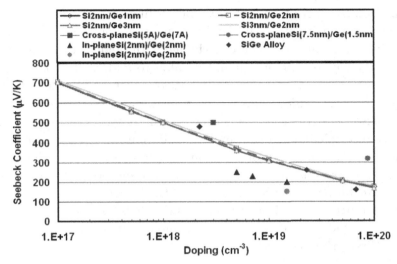

Figure 7.4. Seebeck coefficient of Si/Ge/Si superlattices calculated using the NEGF model compared to experiments.[11–14] Bulusu and Walker, *IEEE Transactions on Electron Devices*, **55**(1), 423–429, (2008).

Figure 7.3 provides a model verification, and we see that the NEGF naturally captures physics that is nonexistent in the CRTA model except through fitting parameters. But how well does the NEGF, with a more first-principles approach, compare to experiment? Figure 7.4 shows the NEGF results next to a handful of experimental measurements from the literature. Not only are the structures in the figure difficult to manufacture with high tolerance, but also the characterization of the structures can be equally challenging. Therefore, high-quality experimental reference data with which to compare are scarce. Nonetheless, the NEGF model seems to match the scattering of experimental data reasonably well.

7.5 First-Principles Calculations

At this point we still cannot calculate thermoelectric properties from fundamental physical properties. Even though we can include noncontinuum effects with the Boltzmann approach and quantum effects with Green's function approach, both models depend on effective material properties. Boltzmann models requires accurate relaxation rates, electron mobility, and confined dispersion, which can change with device structure, and Green's function requires accurate effective mass values, which can change with device structure. True first-principles techniques should be able to address this issue, and many have used density functional theory (DFT) to calculate the band structure, which can be used to estimate electrical conductivity and Seebeck coefficient. Thermal conductivity, on the other hand, is normally calculated using molecular dynamics. Strictly speaking, molecular dynamics is not necessarily a first-principles calculation because it depends on an empirical interatomic potential. However, DFT can be used to provide forces between atoms or can be used as a reference for the development of suitable interatomic potentials. Therefore, for the present discussion, molecular dynamics is considered a first-principles technique.

7.5.1 Molecular dynamics

Molecular dynamics (MD) can be used to calculate the phonon transport in a device with interfaces, complicated crystalline structure, lots of arbitrary boundaries, and so forth. Moreover, its implementation simplicity makes it accessible to even the most naive researcher. In essence, as long as you have good interatomic potentials, which is critical, all you need to do is integrate Newton's law of motion.

$$\sum F = m\vec{a}, \qquad (7.26)$$

where the position and velocity are embedded in the acceleration $(a = dv/dt = d^2x/dt^2)$, and the mass (m) is that of the oscillators (atoms). The forces arise from the chemical potentials between atoms; their generation is beyond the scope of this chapter but can be obtained with electronics structure codes (see below). Fortunately, many excellent open source codes contain prebuilt potentials that are remarkably accurate for well-known crystal systems. LAMMPS[15] is perhaps the most widely used for solids. Once the dynamics have been solved (position and velocity for each atom as a function of time), various averages will reduce to physical quantities.

Three approaches, all of which rely on MD simulation, can be used to learn something about the thermal transport in materials.

The easiest method to conceptualize is nonequilibrium MD (NEMD). In this method, a thermal bias is applied to the molecular system and the amount of energy flowing through the system is recorded. The simulation is set up with two regions of atoms that will be treated as thermal reservoirs. One reservoir is heated by adding energy, and the other reservoir is cooled by removing energy.

Since kinetic energy is related to velocity, the energy is added by velocity scaling. That is,

$$v_{\text{textnew}} = v_{\text{old}} \sqrt{\frac{T_{\text{new}}}{T_{\text{old}}}}, \qquad (7.27)$$

where the amount of energy added is $q = 3/2 N k_B (T_{\text{new}} - T_{\text{old}})$. Each reservoir is held at a different temperature so that the energy leaks from one reservoir and flows into the other reservoir because of the thermal bias. At steady state, the heat flow is related to the imposed temperature difference to obtain the thermal conductivity via Fourier's law,

$$k = \frac{(T_{\text{hot}} - T_{\text{cold}})A}{qL}, \qquad (7.28)$$

where the device area through which the transport occurs (A) and the distance between the reservoirs (L) are known. Criticisms of this method include the following: (1) most MD devices are too small to recover bulk properties, (2) the distribution between the reservoirs is often not linear, which formally invalidates the method, and (3) the heat flux (i.e., the temperature gradient) is far too large to expect the MD simulation to mimic a real system. In reality, each of these criticisms is based in the size argument. Because molecular dynamics tracks the motion of every atom in a system, devices are limited to relatively small sizes, even compared to nanostructures. Consider that a large semiconductor nanocrystal contains something in the neighborhood of 10,000 atoms. While this isolated structure is well within the capability of modern computer clusters, the results of such a simulation are not particularly interesting in a thermal transport point of view for thermoelectric devices. The reason is that the nanoparticle does not represent the transport of a device even if the device is composed of nothing but nanoparticles. Instead most thermal energy is carried by long wavelength modes (small wave-vector) where the group velocity is large. So to capture the thermal performance of thermoelectric materials, we still need to simulate relatively large systems.

Although we can rarely simulate the entire system of interest, we can overcome this limitation by performing thermal conductivity calculations on multiple systems of various sizes. Then the inverse of thermal conductivity is plotted as a function of the inverse of the device size. This linear plot is then extrapolated to $1/l \to 0$ to obtain the infinite size (or bulk) limit of thermal conductivity. We can understand how this works by considering Matthiesen's rule for the lengths that are responsible for phonon scattering. For illustration purposes, we consider only the intrinsic phonon–phonon scattering and boundary scattering. In this case, the thermal conductivity can be expressed as

$$k = cvl, \tag{7.29}$$

where c is the volumetric specific heat, v is the phonon group velocity, and l is the overall mean free path. Now we write the overall mean free path in terms of its constituents,

$$\frac{1}{l} = \frac{1}{l_{p-p}} + \frac{1}{l_b}, \tag{7.30}$$

where l_{p-p} is the phonon–phonon scattering, and l_b is the boundary scattering that is equal to the device size. Now the inverse of thermal conductivity is linearly proportional to the inverse of the device size,

$$\frac{1}{k} = \frac{3}{cv} \left[\frac{1}{l_b} + \frac{1}{l_{p-p}} \right]. \tag{7.31}$$

A fit of the simulated data will provide values for the prefactor and $lp - p$, which yields a bulk value of thermal conductivity to $k = \frac{cvl}{p-p}$. Unfortunately, the prefactor is not always constant and other scattering mechanisms, which give rise to additional mean free paths, often invalidate the method, and care must be taken when applying this method.

Mean free paths often invalidate the method, and care must be taken when applying this method.

The second approach requires equilibrium molecular dynamics. In other words, the simulation is run where the entire device is allowed to remain at a constant average temperature. Without a temperature difference, we cannot directly calculate energy flow. However, energy still sloshes back and forth due to the oscillations of the atoms in the device. To extract conductivity, the Green–Kubo technique requires the autocorrelation of the heat current. The inherent advantage of this technique is that we can impose a periodic system and dispense with the contrived thermal reservoirs as boundaries. In addition, we do not have to verify that the system has reached a steady state or has a linear temperature distribution.

The heat current, however, is not a well-established quantity. It involves (obviously) the oscillation of energy (both kinetic and potential), yet various formulations of the heat current appear to be more or less stable than other formulations depending on the device size, length of time the simulation is run, and complexity of the crystal lattice. An overview of heat current formulations can be found in the literature[16] along with a good discussion on the stability of the autocorrelation for complicated systems.[17]

Another criticism of the equilibrium method is that it requires a large number of time steps, so simulations can be lengthy. In the autocorrelation, a vector that contains the heat current as a function of time is multiplied by itself but with a shifted time index. The shifting must occur over long times. So that the frequency space is sampled appropriately, the simulation must be run for many times longer than the longest time constant. Moreover, it is not always obvious what the longest time constant is.

A third mode of molecular dynamics is less established, so we will coin a name — static lattice methods. These methods are described by imposing a wave packet or phonon mode on an otherwise quiescent device. Another term used to describe the methods is 0 K methods because the lattice starts out without any motion, which implies that the lattice is at absolute zero. From a quiescent state, an artificial wave packet (i.e., displacement and velocity) is imposed, and its propagation is recorded. This method is particularly useful for calculating transmission coefficients at interfaces and reflections off boundaries.

The advantages of the wave packet approach are that the simulations are quick and they yield frequency-dependent quantities. A simulation for a single wave packet is fast because the simulation can be stopped as soon as the wave packet has completely interacted with a structure. The time required to complete the simulation then is the distance you want the phonon to travel (of the order of the device length) divided by the phonon velocity. As in both the equilibrium approach and the nonequilibrium approach, all frequencies and wavelengths should be sampled, so all the phonons should have the chance to bounce back and forth many times and scatter multiple times. However, to obtain frequency information, many simulations must be performed with each at a different wave packet frequency. Therefore, the advantage of fast simulation time is reduced by the necessity to make multiple runs.

7.6 Electronic Structure Methods

Because of the computational complexity of electronic structure calculations, only a very few number of atoms can be considered in any one simulation. Therefore, transport that depends on long-range forces and wavelengths is essentially inaccessible. Nevertheless, density functional theory (DFT) calculations are useful in understanding how the crystal structure, chemical bonding, and valency of atomic species affect the band structure of thermoelectric materials. The band structure can then be used to estimate other properties using kinetic theories. The advantage of this approach is that now all the parameters required to obtain transport ultimately originate from first principles; therefore, heuristics, approximations, and empirical expressions are minimized or even eliminated.

The details of electronic structure calculation are beyond the scope of this chapter. Instead, the reader is referred to the many excellent articles that describe the conduction and valence bands for canonical thermoelectric materials such as bismuth telluride[18] including various alloys, and more structured materials such as cobalt oxides.[19] DFT can also be used to study thermoelectricity in molecular junctions[20] and complex solids such as skutterudites.[21] However, the properties associated with nanostructured materials (superlattices, nanocrystalline composites, etc.) are currently inaccessible with DFT calculations because of the immense computational cost for systems over a few hundred atoms.

References

1. N. W. Ashcroft and N. D. Mermin, *Solid State Physics*. Holt Rinehart & Winston, New York (1976).
2. A. Bulusu and D. G. Walker, Quantum confinement effects on the thermoelectric properties of semi- conductor nanofilms and nanowires. *IEEE Trans. Electron Devices*, **55**(1), 423–429 (2008).
3. S. Datta, *Quantum Transport: Atom to Transistor*. Cambridge University Press: New York, 2005.
4. W. Zhang, T. S. Fisher, and N. Mingo, The atomistic Green's function method: An efficient simulation approach for nanoscale phonon transport. *Numer. Heat Trans. B — Fundamentals*, **51**(4), 333–349 (2007).
5. A. Bulusu and D. G. Walker, Review of transport modeling for thermoelectric materials. *Superlattices Microstructures*, **44**(1–36), (2008).
6. A. Bulusu and D. G. Walker, Modeling of thermoelectric properties of semiconductor thin films with quantum and scattering effects. *J. Heat Trans.*, **129**(4), 492–499 (2007).
7. S. U. S. Choi, Enhancing thermal conductivity of fluids with nanoparticles. In D. A. Siginer and H. P. Wang (eds.), *Developments and Applications of Non-Newtonian Fluids*, pp. 99–105. American Society of Mechanical Engineers: New York, 1995.
8. S. Takagi, J. Koga, and A. Toriumi, Subband structure engineering for performance enhancement of Si MOSFETs. In *Proceedings of the IEEE Electron Devices Meeting*, pp. 219–222, December (1997).
9. T. Wang, T. H. Hsieh, and T. W. Chen, Quantum confinement effects on low-dimensional electron mobility. *J. Appl. Phys.*, **74**(1), 426–430 (1993).
10. V. A. Fonoberov and A. A. Balandin, Giant enhancement of the carrier mobility in silicon nanowires with diamond coating. *Nano Lett.*, **6**(11), 2442–2446 (2006).

11. T. Koga, S. B. Cronin, M. S. Dresselhaus, J. L. Liu, and K. L. Wang, Experimental proof-of-principle investigation of enhanced *Z3DT* in (001) oriented Si/Ge superlattices. *Appl. Phys. Lett.*, **77**(10), 1490–1492 (2000).

12. B. Yang, J. L. Liu, K. L. Wang, and G. Chen, Simultaneous measurements of Seebeck coefficient and thermal conductivity across superlattice. *Appl. Phys. Lett.*, **80**(10), 1758–1760 (2002).

13. B. Yang, J. L. Liu, K. L. Wang, and G. Chen. Cross-plane thermoelectric properties in Si/Ge superlattices. In *MRS Proceedings*, volume 691, pp. G3–2. Cambridge University Press: Cambridge, (2001).

14. W. L. Liu, T. B. Tasciuc, J. L. Liu, K. Taka, K. L. Wang, M. S. Dresselhaus, and G. Chen, In-plane thermoelectric properties of Si/Ge superlattices. In *Proceedings of the 20th International Conference on Thermoelectrics (ICT)*, pp. 340–343, 2001.

15. S. Plimpton. LAMMPS — large-scale atomic/molecular massively parallel simulator. [cited 2015 August 15]; Available from: http://lammps.sandia.gov/

16. D. B. Go and M. Sen, On the condition for thermal rectification using bulk materials. *J. Heat Trans.*, **132**(12), 124502 (2010).

17. A. J. H. McGaughey and M. Kaviany, Phonon transport in molecular dynamics simulations: Formulation and thermal conductivity prediction. *Adv. Heat Trans.*, **39**, 169–255 (2006).

18. S. K. Mishara, S. Satpathy, and O. Jepsen. Electronic structure and thermoelectric properties of bismuth telluride and bismuth selenide. *J. Phys.: Condensed Matter*, **9**, 461–470 (1997).

19. T. Takeuchi, T. Kondo, T. Takami, H. Takahashi, H. Ikuta, U. Mizutani, K. Soda, R. Funahashi, M. Shikano, M. Mikami, S. Tsuda, T. Yokoya, S. Shin, and T. Muro. Contribution of electronic structure to the large thermoelectric power in layered cobalt oxides. *Phys. Rev. B*, **69**, 125410 (2004).

20. P. Reddy, S.-Y. Jang, R. A. Segalman, and A. Majumdar, Thermoelectricity in molecular junctions. *Science*, **315**(5818), 1568–1571 (2007).

21. M. Fornari and D. J. Singh. Electronic structure and thermoelectric prospects of phosphide skutterudites. *Phys. Rev. B*, **59**(15), 9722 (1999).

Chapter 8

Phase-Transition-Enhanced Thermoelectric Performance in Copper Selenide

David R. Brown and G. Jeffrey Snyder

The quality of a thermoelectric material is judged by the size of its temperature-dependent thermoelectric-figure of merit (zT). The ordered phase of Cu_2Se shows a zT peak in its ordered phase, doubling to 0.7 at 406 K over 30 K. The enhancements are due to anomalous increase in their Seebeck coefficients, beyond that predicted by carrier concentration measurements, and band structure modeling. As the Seebeck coefficient is the entropy transported per carrier, this suggests that there is an additional quantity of entropy co-transported with charge carriers. Such co-transport has been previously observed via co-transport of vibrational entropy in bipolaron conductors and spin-state entropy in $Na_xCo_2O_4$. The correlation of the temperature profile of the increases in each material with the nature of their phase transitions indicates that the entropy is associated with the thermodynamcis of ion ordering. This suggests a new mechanism by which high thermoelectric performance may be understood and engineered, and the phenomenological basis of that mechanism is explored.

8.1 Introduction

Thermoelectric device efficiency can be calculated from the operating temperature and three transport variables: electrical conductivity (σ), thermal conductivity (κ) and the Seebeck coefficient (α).[1] To first order, the device efficiency can be calculated from a thermodynamic quantity that is a combination of these transport parameters. The *thermoelectric figure of merit* or zT is expressed as

$$zT = \frac{\alpha^2 \sigma}{\kappa} T. \tag{8.1}$$

The goal of thermoelectric material engineering is to develop materials with larger values of zT. In Eq. (8.1), α and σ are often thought of as purely electronic, while thermal conductivity must include both an electronic component (κ_e) and a lattice vibration component (κ_L) and Eq. (8.1) is rewritten as

$$zT = \frac{\alpha^2 \sigma}{\kappa_e} \frac{1}{1 + \kappa_L/\kappa_e}. \tag{8.2}$$

The two terms in Eq. (8.2) motivate the two principal approaches to improving thermoelectric performance. The first term is typically treated as purely electronic and owing to the reciprocal band structure of the material. It can be optimized by engineering the band structure, which is principally accomplished by altering the carrier concentration through doping, or the band-effective mass or its degeneracy.[2,3] by an intelligent choice of the chemical structure. The second term, containing κ_L, is purely dissipative and great effort is made to reduce it. This is typically done either by adding impurities that lead to preferential scattering of phonons (quanta lattice vibrations) or by finding materials that naturally have low thermal conductivity due to the innate anharmonicities of their structures.[1] For this reason, an ideal thermoelectric material is often referred to as an electron–crystal phonon glass.

In Eq. (8.2) and its general discussion, it is implicit that the thermodynamics of heat and charge transport are all that are relevant to the development of good thermoelectrics. However, in certain materials, there is co-transport of another degree of freedom with charge transport. This leads to an enhancement in the Seebeck coefficient and, therefore, zT. While the Seebeck coefficient is often simply thought of as the voltage induced by a given temperature gradient (i.e., $\alpha = \frac{\nabla V}{\nabla T}$), it can also be expressed in terms of the entropy transported per electron[4-6]:

$$\alpha = -\frac{S^*}{q}. \tag{8.3}$$

By convention, the sign in Eq. (8.3) is negative. It ensures that p-type materials have a positive Seebeck and n-type materials have a negative Seebeck. Equation (8.3) states that the Seebeck coefficient is equal to the entropy transported per charge; however, it does not specify that this entropy must be *electronic* entropy. Co-transport of nonelectronic entropy can cause increased Seebeck and, therefore, increased zT beyond that specified by band structure calculations. Denoting the electron degree of freedom only Seebeck as α_e and the co-transported Seebeck as $\Delta\alpha$, Eq. (8.2) may be reformulated as

$$zT = \frac{\alpha_e^2 \sigma}{\kappa_e} \frac{1 + (\Delta\alpha/\alpha_e)^2}{1 + \kappa_L/\kappa_e}. \tag{8.4}$$

Such entropy co-transport has previously been observed in three different types of material systems: vibrational entropy co-transport has been observed in boron carbide,[7] lattice spin entropy co-transport has been observed in Na_xCoO_2,[8,9] and lattice entropy co-transport has been observed in phonon-drag systems.[10] In the case of Na_xCoO_2, for example, the differing spin degeneracy of electron-occupied and electron-unoccupied cobalt sites provides the mechanism for this coupling of carrier transport to entropy transport.[11] However, this strategy has, thus, far been limited to small changes in spin degrees of freedom of single ions; it remains an open question whether structures with more spin degrees of freedom can be coupled to charge transport.

The focus of the discussion of this chapter is on the recently observed Seebeck and zT peak observed by these authors in Cu_2Se.[12] This peak is observed near the materials $\approx 410\,K$ phase transition. We suggested that this peak is due to co-transport of structural entropy associated with the disordering of that material's super-ionic phase transition. Here we consider coupling the carrier transport to degrees of freedom associated with the entropy associated with an order–disorder phase transition.

A phase transition is always associated with an entropy change because there is always a concurrent transformation in system symmetries.[13] If the entropy change of a continuous phase transition can be associated with carrier transport, a substantial enhancement in Seebeck may be obtainable. The number of degrees of freedom associated with a structural transformation scales as the number of atoms in the system rather than the number of carriers. For a typical thermoelectric material with a carrier concentration of 10^{20} cm^{-3}, there are 100 times as many atoms as there are charge carriers. The entropy change of a super-ionic phase transition is similar to that of melting.[14] Thus, the potential Seebeck enhancement by this mechanism may be extremely large. Because phase transitions occur at a discrete temperature, it is relatively simple to distinguish the anomalous enhancement due to ordering entropy co-transport from the band structure contribution. In a material without a phase transition, such enhancements may be misattributed to the band structure through incorrect determination of one of the band parameters.

8.2 Lambda-like Phase Transition of Cu$_2$Se

All previous authors considered the ion-disordering transition of Cu$_2$Se to be a first-order phase transition.[14–18] The work presented here shows it to be definitively of second order.[12] This question is of central importance; the zT of Cu$_2$Se cannot be properly calculated without understanding the nature of its phase transition. The determination of the differential scanning calorimetry (DSC) data depends on the order of the phase transition. A substantial broad peak is seen in the DSC for Cu$_2$Se; see Fig. 8.1. If the transition is of second order, the DSC measurement must be treated as c_p. If it were of first order, it would be more proper to use the Dulong–Petit heat capacity instead.

As κ is calculated from c_p by the formula $\kappa = \rho c_p D_T$, and zT from κ by $zT = \alpha^2 \sigma T / \kappa$, this argument has an order of magnitude impact on the zT calculated. Treating Cu$_2$Se as a first-order transition results in a fivefold overestimate in zT above its true value at

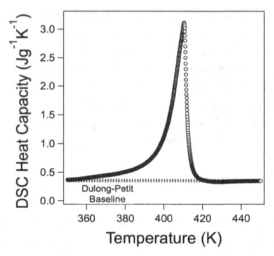

Figure 8.1. Differential scanning calorimetry c_p for Cu_2Se (blue circles) with the Dulong–Petit contribution as a green dotted line. Determining the order of the phase transition of Cu_2Se determines which of these two curves should be used to calculate c_p. Adapted from data presented in Ref. [12].

the 406 K peak. Answering this question is particularly pressing due to recent published articles,[17,18] claiming zT two to five times that discussed here based on the first-order treatment of their calorimetry data.

There are two principal types of order–disorder phase transitions. First-order transitions are characterized by a first derivative discontinuity in a thermodynamic state parameter. This is typically measured by the instantaneous change in the volume or enthalpy as a thermodynamic parameter is changed.[13] A second-order transition shows a discontinuity in a second or higher order derivative of the thermodynamic quantities.

Certain second-order transitions, including that of Cu_2Se, have a derivative of free energy that diverges to infinity at the phase transition temperature. Under the old classification system of Ehrenfest, these were not considered second-order transitions.[19] Instead, they were called λ-transitions for the characteristic λ-shape in the dependence of their heat capacity on temperature.[20,21] In more modern classification systems based on the work of Landau and Ginzburg, all phase transitions with continuous transformation of free energy are considered to be of second order or continuous.[22]

In proximity to an order–disorder phase transition, the free energy is described in terms of an order parameter. For example, in the canonical case of a spin-ordering transition, the order parameter is the average spin magnetization. The super-ionic-order parameter (m) is theorized to relate to the relative population of different sites by the mobile ion.[23] For a second-order transition, the Landau Free Energy is

$$F = am^2 + bm^4. \tag{8.5}$$

As the phase transition is approached, thermodynamic quantities follow critical power laws. For example, the order parameter decreases to zero continuously (Fig. 8.2):

$$m = m_0 \left(\frac{T_c - T}{T_c}\right)^{\beta} \tag{8.6}$$

in which β is called the critical exponent. Order parameter in a structural transformation is related to the diffractogram peaks, so

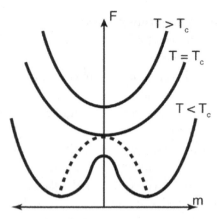

Figure 8.2. Free energy (F) versus order parameter (m) for a second-order transition. Each of the curves is on its separate axis. The dotted line represents the equilibrium order parameter. As T goes to T_c, m goes continuously to zero.

a related power law should be discoverable by analyzing those peaks. A similar power law is expected for heat capacity.

The description of the phase transition behavior of Cu_2Se is complicated by the unsettled argument over the nature of that phase transition and the structure of the low temperature phase. At this point, the high-temperature phase is fairly well understood.[26] All the compositions of $Cu_{2-\delta}$ Se from $\delta = 0$ to $\delta = 0.2$ appear to have the same high-temperature anti-fluorite cubic structure;[24,25,27] see Fig. 8.3. Although that phase is on average anti-fluorite, significant Cu^+ occupation of trigonal planar and octahedral interstices is observed.[27] Hopping through these interstices is the mechanism of fast copper ion transport; the ion transport pathways have been successfully determined to be along the [1 1 1] direction from tetrahedral to trigonal planar intersticies.[28]

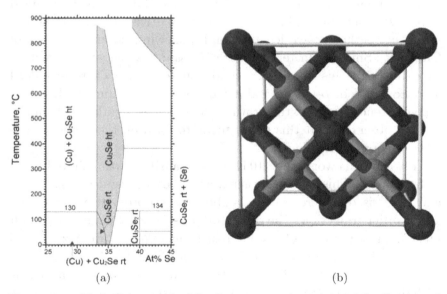

(a)   (b)

Figure 8.3. Phase diagram (a) of Cu–Se system in the vicinity of Cu_2Se adapted from Heyding[24] and the ASM Phase Diagram. The phase transition is between the β-Cu_2Se(RT) phase and the α-Cu_2Se}(ht) phase. Anti-fluorite structure (b) of which α-Cu_2Se is a modification. Se is coordinated FCC and is represented in red. Ground state Cu is tetrahedral coordinated to the Se though significant occupation of trigonal planar and octahedral interstitials has been measured. The structure of β-Cu_2Se is unknown.

Despite Cu_2Se presence in a mineral form and its binary composition, the ordered low-temperature structure is yet unknown. This is not for the lack of trying. In 1987, Milat *et al.*[30] proposed a monoclinic supercell and in the course of that work noted eleven other proposed structures. They proposed a structure that assumed significant octahedral occupation. Later authors proposed more complex superstructures.[18,29] These structures are insufficiently complex to explain the crystallography data presented below.

Multiple authors have proposed a co-existence transformation between the α and β phases with a temperature width of tens of kelvins.[15,17] This hypothesis, though reasonable, is contradicted by the data presented in this chapter. There are three reasons for this. The $Cu_{1.8}Se$ is commensurate with the α Cu_2Se. In the region $\delta = 0.05$ to $\delta = 0.2$, there is actually a co-existence of the β-Cu_2Se phase and the $Cu_{1.8}Se$; if the phase diagram is determined imprecisely, the single-phase region goes unnoticed. Finally, unknown errors in synthesis have led to samples showing impurity phases of $Cu_{1.8}Se$.[17] Since room temperature $Cu_{1.8}Se$ has the same structure as α-Cu_2Se,[24] this is an easy confusion to make. There is indeed a co-existence of β-Cu_2Se and α-$Cu_{1.8}Se$ in such samples, but it is a co-existence of admixture rather than that of synthesis. Vengalis *et al.*[31] observed that this phase tends to form on the grain boundaries of copper-rich phases.

Prior to this work, the 410 K phase transition was believed to be of first order. This is unsurprising, as it takes careful measurement and analysis to differentiate a lambda second-order transition from a first-order transition. The difficulties of this determination are well illustrated in the case of the lambda transition of β quartz.[32,33] As late as 1980, authors were still confused about the lambda nature of its phase transition.[33] While structural second-order transitions with diverging heat capacity are of interest to the physics community, they are far less common than the first-order transitions. Korzhuev determined Cu_2Se's transition to be of first order on the basis of the Clausius–Clapeyron relations.[34] However, Pippard showed than an analogous relation holds for lambda-type transitions.[21] Vučić determined it to be of first order on the basis of its sharp feature in their dilatometry data;[16] again, such sharp features are also expected in the case of a lambda-type second-order transition.[21] Qualitative assessment of sharpness of thermophysical peak at a phase transition temperature can differentiate a first-order transition from a second-order transition without diverging heat capacity;[35] it is insufficient for differentiating a first-order transition from a lambda-type second-order transition.

Despite this confusion, there is some certainty about the phase diagram at room temperature. By electrochemical determination.[36] Korzhuev *et al.* found there to be a single-phase region of $Cu_{2-\delta}Se$[37] for $\delta = 0$ to $\delta = 0.05$ and a range of co-existence of $Cu_{1.95}Se$ and $Cu_{1.8}Se$ from $\delta = 0.05$ to $\delta = 0.20$. Temperature-dependent dilatometry was performed by Vučić *et al.* in the single-phase region, and he developed a phase diagram on this basis; see Fig. 8.4. Vučić's collaboration made follow up transport measurements.[25,30,38,39] In general, our sample shows the properties that Vučić observed in his samples of nominal composition $Cu_{1.99}Se$.

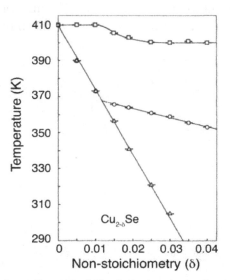

Figure 8.4. The phase diagram of Cu_2Se in its single phase region by Vučić.[25] This diagram was established by dilatometry. Notably, there are multiple phase transitions. Reprinted figure with permission from Ref. [35]. Copyright 1981 by the American Physical Society.

8.2.1 Diffractometry

A diffractogram was measured at room temperature from $2\theta = 10°$ to $2\theta = 90°$. The sample was then heated at 1 K per minute to 425 K (above the nominal phase transition temperature of 410 K. A second diffractogram was taken at 425 K. These diffractograms are shown in Fig. 8.5. The low-temperature diffractogram is consistent with literature reports.[29] The bifurcation of the major α-Cu_2Se peaks in the β-Cu_2Se phase is consistent with the β phase being a monoclinic or orthorhombic modification of the anti-fluorite structure. $Cu_{1.8}Se$ impurity phase observed by Liu *et al.* was not observed here.[17]

For the crystallographic determination of the nature of the phase transition, diffractometry must be performed at a series of temperatures that transverse that transition. Diffractograms were continuously measured on heating from $2\theta = 23°$ to $2\theta = 45°$. The duration of each scan was three minutes and consequently the temperature changed by 3 K from start to end of each scan. The 2θ range was chosen because of the excellent signal intensity and the two separate bifurcated peaks seen. Visualization of these data (Fig. 8.6) shows a continuous evolution of the bifurcated peaks at

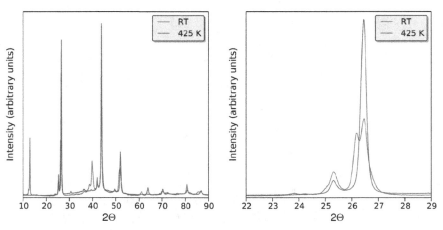

Figure 8.5. PXRD of Cu_2Se at 300 K and 425 K from $2\theta = 10°$ to 90° (left). Zoom in near the 26° peak set(right). Peaks' positions as identified in literature were observed.[29] The sample is single phase. Adapted from data presented in Ref. [12].

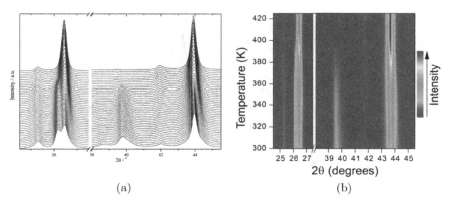

Figure 8.6. Temperature varied diffractograms of Cu_2Se. Data are presented as stacked diffractograms (a) and as a color map (b). The peak intensities and angles shift continuously from the low-temperature to the high-temperature phase. Figures adapted from data presented in Ref. [12].

low temperature into the single peaks at high temperature. Both the peak intensities and angles shift continuously from the low- to high-temperature phase. This strongly contrasts with the abrupt change that would be seen as in a first-order transition.

The continuous phase transition is further confirmed by the pair distribution function (p.d.f.) analysis of total scattering data; see Fig. 8.7. The changes of the p.d.f. are gradual, indicating that the ordering of Cu interstitials occurs over a wide temperature range. There is no evidence of a first-order discontinuity in peak positions or of the β phase being present below 410 K. The phase transition does not appear to be complete until 450 K; transformation above the phase transition temperature is the characteristic of second-order transitions. The $Q_{\max} = 26$ Å used is insufficient for truly accurate quantitative fitting, as indicated by the presence of substantial integration error ripples below the first peak maxima.

Even without modeling the data, it is possible to extract qualitative information. By studying the high-temperature structure of Cu_2Se, it is clear that the peak at 4.1 Å (Fig. 8.7(a)) is a superposition of the shortest Cu–Cu and Se–Se distances in the [1 1 0] direction. Above 300 K, the peak becomes increasingly asymmetric, indicative of multiple Cu–Cu distances in the high-temperature phase

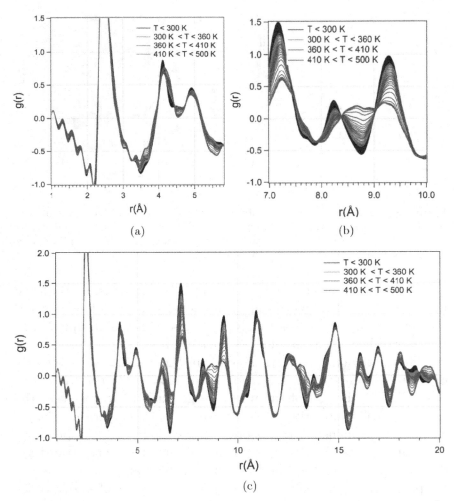

Figure 8.7. Pair distribution function data for Cu$_2$Se. The unit cell size is 5.8 Å. The coordination number remains the same through the phase transition (a) but correlations between high temperature equivalent unit cells breaks down. (b) Full data set. (c) Figures previously presented in Ref. [12].

related to the disorder of Cu interstitials. At low temperature, the Cu order to form a superstructure. The superstructure formation is most clearly seen in the region from 8 Å to 9.5 Å (Fig. 8.7(b)). This range corresponds to Cu–Cu distances in the [1 1 0] direction in adjacent cubic unit cells. Below 410 K, there are two distinct peaks at 8.2 Å and 9.3 Å. However, in the high-temperature phase, the same region is a continuum of overlapping peaks arising from the disorder of Cu.

8.3 Transport Near the Phase Transition of Cu$_2$Se

Unusual transport effects have been observed before near the critical temperature in Cu$_2$Se;[18,40] however, this is the first work that also considers changes in all thermal transport measurements (D_t, κ, c_p) to derive an improved value for zT. Liu *et al.* report a zT greater than 1;[17] however, their work assumes all the heat released as measured by DSC is due to the first-order structural transformation. In that case, it would be appropriate, as is done by Liu *et al.*, to calculate κ and zT using the smaller Delong–Petit heat capacity. Due to the continuous nature of the 410 K super-ionic phase transition, the elevated peak in calorimetry is an equilibrium rather than the kinetic aspect of the system behavior and must be used to determine zT.

The study of electrical transport near the phase transition of Cu$_{2-\delta}$Se owes mostly to the work of Vučić and his collaborators at the University of Zagreb.[16,25,30,38,39,41] His measurements spanned from the $\delta = 0$ to $\delta = 0.045$ single-phase range.[42] Based on dilato-metric measurements, he developed a phase diagram for Cu$_{2-\delta}$Se; see Fig. 8.4. At all compositions, he found a phase transition at $\approx$410 K and a second-phase transition at a lower temperature with temperature dependent on δ. The data presented on the sample of Cu$_2$Se presented here correspond generally to his observations of the $\delta = 0.01$ sample. WDS data on Cu$_2$Se bounded $\delta < 0.005$ for this sample; the reason for this disagreement is unclear.

Electrical conductivity (Fig. 8(a)) was measured at a heating rate of 10 K h^{-1}. It shows three main features: a knee at 355 K, a minimum at 400 K, and a kink at 410 K. The kink at 410 K corresponds to the observed phase transition in the crystallography data. In general, shape $\sigma(T)$ strongly resembles the data in Vučić's studies,[39] though it corresponds best to his Cu$_{1.99}$Se stoichiometric sample.

The knee in conductivity at $T_{c2} = 355$ K corresponds to the lower temperature second-order phase transition measured by Vučić.[39]

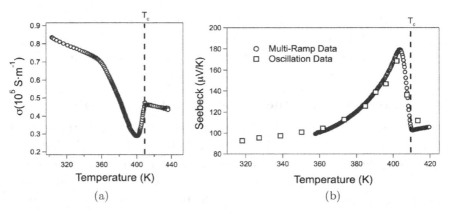

Figure 8.8. Electrical conductivity (a) and Seebeck coefficient (b) measured through the 410 K phase transition. Adapted from data presented in Ref. [12].

He empirically determined a power law for σ below T_{c2}:

$$\sigma = \sigma_0 \left(1 - \frac{T}{T_{c2}}\right). \qquad (8.7)$$

From the low-temperature slope, a predicted temperature $T_c = 360\,\mathrm{K}$, was determined, which is consistent with the T_{c2} determined by inspection. Above the phase transition, the conductivity is again linear, though 20% lower than the value predicted by extrapolation from the low-temperature behavior.

For improved accuracy and temperature resolution, the Seebeck was measured through its phase transition by a new methodology which is described in Appendix A. Data from both the standard "Oscillation" method and the new "Multi-Ramp" method are plotted in Fig. 8(b). The three significant features observed in σ are echoed in α. There is a kink in the Seebeck at 410 K, corresponding to the main phase transition observed in crystallography. Below 360 K, α is a linear function of temperature; above it shows nonlinear behavior. There is a maximum in α at 403 K, at a slightly elevated temperature compared to the 400 K minima in electrical conductivity. Above the phase transition, the Seebeck is locally linear, though 10% lower than the value predicted by the extension of the low-temperature trend.

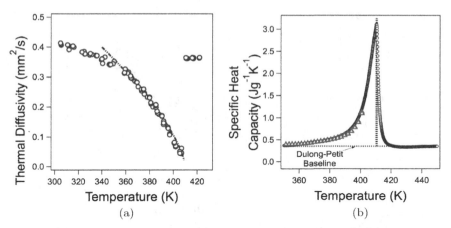

Figure 8.9. Thermal diffusivity (a) and specific heat capacity (b) for Cu$_2$Se from its room temperature through its 410 K phase transition. Adapted from data presented in Ref. [12].

Thermal conductivity was calculated from measurements of density (ρ), DSC heat capacity (C_p), and laser flash diffusivity (D_T). Density was measured to be 6.7 g cm^{-3} by geometric calculation and confirmed by the principle of Archimedes. Thermal diffusivity decreases linearly from 300 K to 360 K; see Fig. 8.9(a). From 360 K to 410 K, the thermal diffusivity shows an excellent fit to a critical power law with $T_c = 410$ K and critical exponent $r = 0.80$. Above the 410 K phase transition, the thermal diffusivity is again changes only in a steady linear fashion with temperature.

Heat capacity is presented here in Fig. 8.9(b), with the inclusion of quasi-static data as measured by a Quantum Design PPMS. Below 360 K and above 425 K, the heat capacity gives a baseline value of 0.374 J g^{-1} K^{-1}, consistent with the Dulong–Petit c_p for Cu$_2$Se, 0.361 J g^{-1} K^{-1}. The 60 K breadth of the peak is indicative of the continuous nature of the transition. In the transition region, there is a lambda-type peak, as is the characteristic of continuous phase transitions in ionic conductors.[43]

From the transport properties aforementioned, zT was calculated (Fig. 8.10). zT doubles over a 30 K range peaking at 0.7 at 406 K. Although strong nonlinearity in each of the individual transport

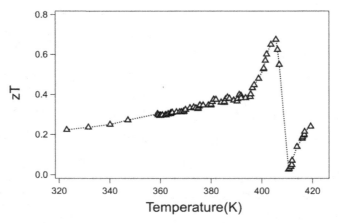

Figure 8.10. zT for Cu_2Se from its room temperature through its $410\,K$ phase transition. Adapted from data presented in Ref. [12].

variables onsets at $360\,K$, there is no nonlinear in zT until $390\,K$. This suggests that more than different effects — perhaps associated with the multiple Cu_2Se phase transitions — are needed to explain the anomalous transport behavior between $360\,K$ and $410\,K$.

8.3.1 Band Structure Analysis

The electronic transport properties of heavily doped thermoelectrics can be typically described by modeling their electronic band structures.[1,44] If the full electronic band structure is known — or more realistically predicted by density functional theory — then the transport coefficients could be computationally determined (e.g., by BoltzTRaP[44]). However, simplified models that take into account only part of the band structure are extremely successful at predicting transport properties.[3] These models are successful because only band states within $3k_bT$ of the electron chemical potential (i.e., the Fermi level) contribute significantly to electron transport.[2] The starting point for these models is the single parabolic band (SPB).[45] Thermoelectrics are heavily but not metallically doped, so that the band of the dominant conductor tends to dominate, but the Fermi level is not far from the band edge. An SPB has a quadratic dependence of energy (E) on the reciprocal lattice vector (K) relationship of the form

$$E = \frac{\hbar^2 (k - k_0)^2}{2m^*}. \tag{8.8}$$

If the effective mass (m^*) and the band chemical potential (μ) are known, then the carrier concentration may be written as

$$n = 4\pi \left(\frac{2m^* k_b T}{h^2} \right)^{\frac{3}{2}} F_{\frac{1}{2}}(\eta) \tag{8.9}$$

in which h is Planck's constant, $\eta = \frac{\mu}{k_b T}$ is the reduced chemical potential, and $F_j(\eta)$ is the Fermi integral of order j.

If the energy dependence of scattering (λ) is also known, then transport variables may be modeled as well. In the case of scattering by acoustic phonons, $\lambda = 0$. This is a good assumption for thermo-electric materials above the Debye temperature. In this model, the Seebeck coefficient may be expressed as

$$\alpha = \frac{k_b}{e} \left(\frac{(2 + \lambda) F_{\lambda+1}}{(1 + \lambda) F_\lambda} - \eta \right). \tag{8.10}$$

The general behavior can be understood well if the degenerate (e.g., metallic) limit of Eq. (8.10) is taken:

$$\alpha = \frac{\pi^{\frac{8}{3}} k_b^2}{3qh^2} n^{\frac{-2}{3}} T m^* (1 + \lambda). \tag{8.11}$$

Except for the region of elevated zT (390–410 K), all the observed variations in transport can be explained by a simple band structure model. From the knowledge of the band structure, the energy dependence of scattering, and the reduced chemical potential, all thermoelectric transport properties can be modeled for typical systems. Although η cannot be easily measured, it can be inferred from the Seebeck coefficient from Eq. (8.9). In the degenerate (metallic) limit, this dependence can be expressed in a simple closed form

$$n = 4\pi \left(\frac{2m^* kT}{h^2} \right)^{\frac{3}{2}} F_{\frac{1}{2}}. \tag{8.12}$$

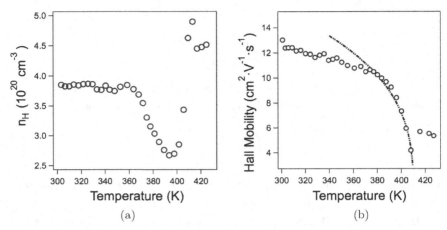

Figure 8.11. Hall carrier concentration (a) and Hall mobility (b) of Cu_2Se. The minimum of n_H is at 390 K, while μ_H decreases until 410 K. μ_H could be fit to a power law with a critical exponent $4 = 0.32$. Adapted from data presented in Ref. [12].

The Hall coefficient (R_H) and electrical conductivity were measured via the Van der Pauw method using a 1 Tesla magnetic field at the NASA-JPL thermoelectrics laboratory. From these measurements, n_H and the Hall carrier mobility were extracted; see Fig. 8.11. From 300 K to 360 K, n_H is constant, while μ_H decreases steadily. The linear decrease in σ (Fig. 8(a)) observed is entirely due to a decrease in mobility. Given the association with the copper disordering phase transition, it is possible that a scattering mode like the dumb-bell mode of Zn_4Sb_3 is steadily activated in this temperature range.[46] Between 360 K and 410 K, the Hall carrier concentration dips until it reaches a minimum of 2.7×10^{20} cm^{-3}. This minimum occurs at 393 K, 10 K lower than the minimum in electrical conductivity and the maximum in the Seebeck coefficient. This minimum is also very close in temperature to where the zT shows its nonlinear increase in temperature, suggesting that the mechanism that causes the increase in zT also causes the change in the trend of n_H. During this temperature range, μ_H could be fit to a power law with a critical exponent $r = 0.32$. The Hall mobility remains low in the high-temperature phase with a value 30% below that expected from the extrapolation of low-temperature behavior.

Equation (8.12) suggests that three factors may cause an anomalous increase in the Seebeck coefficient: a decrease in the carrier concentration (n_H), an increase in the scattering parameter (λ), or an increase in the band effective mass (m^*). An increase in λ is unlikely. The structural delocalization as presented in our p.d.f. data (Fig. 8.7) may lead to an increase in the intensity of acoustic phonon scattering, but it will not alter the effect's energy dependence. Near a phase transition, a low-frequency optical phonon mode — a *Goldstone* mode — may be present. However, the energy dependence of optical phonon scattering via lattice deformation is the same as that for electrons, and the dependence via dipole effects is only slightly larger than that for acoustic phonons.[47]

Up to $380\,\text{K}$ the data can be explained entirely with an SPB with $m^* = 2.3m_e(\pm 5\%)$. Above this temperature, an increase in m^* up to 50% is required followed by an even more sudden decrease. This is physically inconsistent with the continuous transformation observed via crystallography. Effective mass can be related to the band structure at the Fermi level by the formula

$$m^* = \hbar^2 \left(\frac{\partial^2 E}{\partial k^2} \right)^{-1}. \tag{8.13}$$

Equation (8.13) means that a substantial change in m^* requires a substantial change in the reciprocal space band structure. As the reciprocal space band structure is related by Fourier transformation to the physical spatial representation of the atomic orbitals and, thus, the coordination of the atoms, a substantial change in m^* would, therefore, require a significant change in local electron coordination — a change that is inconsistent with the minor change in the band structure seen in the temperature-resolve PXRD. It would further require this change to be transient, existing only in proximity to the $410\,\text{K}$. However, the electron bands are dependent on the average structure rather than the correlation length of the order parameter. Thus, we expect that the effective mass and other band structure attributes should vary smoothly from one phase to the other, rather than peaking at the phase transition.

While the observed shift in carrier concentration cannot explain the peak in Seebeck, it does elucidate one apparent anomaly in the

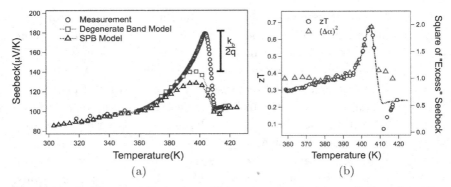

Figure 8.12. Measured Seebeck coefficient compared with predictions from the band structure model with $m^* = 2.3m_e$ and measured n_H (a). The square of the Seebeck "excess" to the band structure prediction (b) explains the zT peaks' size and breadth. Adapted from data presented in Ref. [12].

transport data. As discussed in the previous section, the onset of non-linear transport behavior occurs at 360 K, but the zT shows a visible deviation from a linear trend only at 393 K. The measured Seebeck coefficient and that predicted by Eq. (8.11) and an SPB model with $m^* = 2.3m_e$ are compared in Fig. 8.12. Near 385 K, both models show an increasing deviation from the measured data. This increase is of the order of the natural scale of Seebeck, $k_b/2q = 43\,\mu V^{-1}\,K^{-1}$; the increase corresponds to the transport of the entropy of an additional degree of freedom per electron.

When the square of the measured Seebeck divided by the band structure predicted Seebeck is compared with zT, as in Fig. 8.12(b), it is seen that the anomalous increase in Seebeck almost explains the observed breadth and height of the zT peak. The measured Seebeck is 48% higher than the prediction of the SPB model and 40% higher than the prediction of the degenerate band model at the temperature of peak zT, 406 K. The measured zT is 60% higher at its 406 K peak compared to linear extrapolation from its increase from 360 to 385 K. The "excess" Seebeck ($\Delta\alpha$) compared to the band structure slightly overestimates the height of the zT peak. Both the zT and $\Delta\alpha$ are increased noticeably over the exact same temperature range of 393–410 K. This suggests that some aspect of the lambda-type phase transition increases the zT of Cu_2Se.

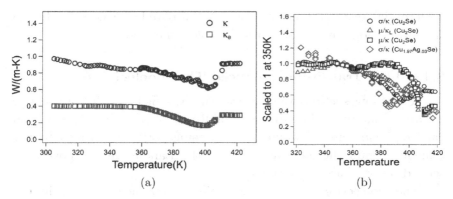

Figure 8.13. Decreasing thermal conductivity does not cause the zT peak. (a) The decrease in κ is due to the electronic contribution (κ_e). (b) Electrical properties decrease slower than thermal properties in the phase transition region and thereby diminishing the zT peak. Adapted from data presented in Ref. [12].

The zT increase cannot be explained by a relative improvement of phonon to electron scattering, that is, by the σ/κ contribution to zT. While there is insufficient data to truly determine the Lorenz number (L) over the entire phase transition region, single temperature estimates bound it between $1.8\,\mathrm{W}\,\Omega^{-1}\,\mathrm{K}^{-2}$ and $2.0\,\mathrm{W}\,\Omega^{-1}\,\mathrm{K}^{-2}$. This allows the estimation of κ_e by the formula $\kappa_e = L\sigma T$. The estimated electronic portion of the thermal conductivity qualitatively explains the observed decrease in total thermal conductivity; see Fig. 8.13(a).

A direct comparison of the electron (μ_H, σ) and thermal transport (κ, κ_L) indicates that the zT is not increased by preferential scattering of phonons over electrons; see Fig. 8.13(b). Although thermal conductivity decreases between $360\,\mathrm{K}$ and $410\,\mathrm{K}$, this decrease is more than counteracted by a decrease in electrical mobility over the same range. The observed increase in zT is not due to the reduction of the thermal conductivity via preferential scattering of phonons over electrons. This trend is particularly clear when comparison is made with κ_L instead of κ. This suggests that across the entire temperature range κ_L is close to its glass-like minimum, such that further increased scattering from thermally activated modes associated with ion disordering cannot reduce it significantly further. This contradicts the proposal of Liu *et al.* that the increase in zT is due to preferential scattering of phonons as compared to electrons due to the interaction with a soft optical (Goldstone) mode.[18]

8.4 Phase-Transition-Enhanced Thermoelectrics

In the previous section, we have seen that Cu_2Se has a sharply peaked enhancement in α that drives a doubling of zT. This observation can be phenomenologically explained using both the Onsager theory nonequilibrium thermodynamics and the Landau theory governing second-order phase transitions.

Thermoelectric transport coefficients can be expressed in the Onsager coefficient formalism. As typically presented[6] for a thermoelectric system, this is,

$$\begin{bmatrix} J_e \\ J_s \end{bmatrix} = - \begin{bmatrix} {}^2L_{ee} & {}^2L_{es} \\ {}^2L_{se} & {}^2L_{ss} \end{bmatrix} \begin{bmatrix} \nabla \tilde{\mu}_e \\ \nabla T \end{bmatrix}. \tag{8.14}$$

The subscript 2 is used here for didactic purposes. Note that $V = \frac{\tilde{\mu}_e}{e}$ and J_e is a quantity flux rather than a charge flux, i.e., defined by $J = q J_e$. The Onsager matrix in Eq. (8.14) is derived from a two-term entropy production equation of the form[5]

$$T\dot{S} = -J_s \cdot \nabla T - J_e \cdot \nabla \mu_e. \tag{8.15}$$

But imagine that if there is a second thermodynamic quantity that transports — without loss of generality, let us call it m and its conjugate force h — then the entropy production equation would be

$$T\dot{S} = -J_s \cdot \nabla T - J_e \cdot \nabla \mu_e - J_m \nabla h. \tag{8.16}$$

If $J_m > 0$, then Eq. (8.16) will apparently not capture the entirety of the entropy transported, though this problem can be overcome if J_s is replaced with some other entropy-like flux.[5] In this case, there would be dissipation from a nonelectronic transport quantity that must affect the maximal zT and η of the system. In the analysis in the following, we principally consider the condition of $J_m = 0$, and so this complication does not enter in.

The Onsager matrix that corresponds to Eq. (8.16) is

$$
\begin{bmatrix} J_e \\ J_m \\ J_s \end{bmatrix} = - \begin{bmatrix} {}^3L_{ee} & {}^3L_{ms} & {}^3L_{es} \\ {}^3L_{me} & {}^3L_{mm} & {}^3L_{ms} \\ {}^3L_{se} & {}^3L_{sm} & {}^3L_{ss} \end{bmatrix} \begin{bmatrix} \nabla\tilde{\mu}_e \\ \nabla h \\ \nabla T \end{bmatrix}. \tag{8.17}
$$

The essential complication arises from this: ${}^2L_{se} \neq {}^3L_{se}$ for all conditions, but ${}^3L_{se}$ is calculated by the band structure models presented earlier and generally calculated from density functional theory. The discrepancy between them gives the contribution of entropy co-transport to thermoelectric performance. This is not to say that the kinetic theory that underlies such calculations is incorrect, but rather that if incomplete information is given to them, they cannot provide the correct results. The measurement of the Seebeck coefficient is always given by

$$
\alpha = \frac{\nabla V}{\nabla T}\bigg|_{J_e=0} = \frac{1}{q}\frac{-{}^2L_{se}}{{}^2L_{ee}}, \tag{8.18}
$$

with $J_e = 0$ indicating that the material is measured with an open electrical circuit. The relation between the Onsager L coefficients from Eq. (8.15) and those from Eq. (8.16) will depend on the boundary conditions, i.e., $\nabla h = 0$ or $\nabla m = 0$. The cases of both ∇h, $\nabla\widetilde{\mu_e} = 0$ and $J_m = 0$, $J_e = 0$ were considered by DeGroot in the context of thermodiffusion,[5] while the case of $J_e = 0$, $\nabla m = 0$ was recently consider by Sandbakk *et al.*[48] in the context of coupled ion and volume transport in ion–membrane thermoelectrics.

In the case of $\nabla h = 0$, the Seebeck expressed in terms of Eq. (8.17) is

$$
\alpha = \frac{\nabla V}{\nabla T}\bigg|_{J_e=0, \nabla h=0} = \frac{-1}{q}\frac{{}^3L_{se}}{{}^3L_{ee}}. \tag{8.19}
$$

This gives the same form as that of Eq. (8.18), except that there may be some external dissipation due to $J_m \neq 0$ that limits the heat to electron conversion efficiency possible. The conditions of a normal thermoelectric may be thought of as having $L_{mm} = 0$, so that $J_m = 0$ implies $\nabla h = 0$ and there is, therefore, neither co-transport nor dissipation. The other transport coefficients (σ, κ) will behave similarly.[48]

In the case of $J_m = 0$, the Seebeck expressed in terms of Eq. (8.17) is

$$\alpha = \frac{\nabla V}{\nabla T}_{J_e=0,J_m=0} = \frac{-1}{q} \frac{{}^3 L_{se}}{{}^3 L_{ee}} \left(\frac{1 - \frac{{}^3 L_{ms}}{{}^3 L_{mm}} \frac{{}^3 L_{me}}{{}^3 L_{es}}}{1 - \frac{{}^3 L_{me}}{{}^3 L_{mm}} \frac{{}^3 L_{me}}{{}^3 L_{ee}}} \right). \tag{8.20}$$

The expression $\frac{{}^3 L_{ms}}{{}^3 L_{mm}}$ is equivalent to the Seebeck coefficient but associated with the transport of m instead of electron transport, that is,

$$\alpha_m \equiv \left(\frac{\nabla h}{\nabla T} \right)_{\nabla \tilde{\mu}_e, J_m=0} = \frac{{}^3 L_{ms}}{{}^3 L_{mm}}. \tag{8.21}$$

Like α, α_m has its own presence contribution and that may be defined as

$$\alpha_{m,\text{presence}} = - \left(\frac{\partial s}{\partial m} \right)_{u,n}. \tag{8.22}$$

The expression $\frac{{}^3 L_{em}}{{}^3 L_{ee}}$ is the co-transport of J_m with J_e when there is no driving force for the direct transport of m. It will have its own presence contribution:

$$\left(\frac{J_m}{J_e} \right)_{\nabla T=0, \nabla h=0} \approx \left(\frac{\partial m}{\partial n} \right)_{u,s}. \tag{8.23}$$

This expression will be most accurate if all transport of m is mediated by the transport of the electrons. This is certainly the case in the spin-state enhancement of Seebeck observed in oxide thermoelectrics.[9,49] Due to the independent mobility of ions in Cu_2Se, Eq. (8.23) is here only an approximation.

By combining Eqs. (8.22) and (8.23) into a single expression, a quasi-thermodynamics expression for the entropy co-transport Seebeck may be obtained. Let us call this term $\alpha_{\text{order-entropy}}$. It can be formulated in terms of the free energy density (f) and T as

$$\alpha_{\text{order-entropy}} = \frac{-1}{q} \left(\frac{\partial f}{\partial m}\right)_{T,n} \left(\frac{\partial m}{\partial n}\right)_{T,s}, \tag{8.24}$$

with f the Landau free energy (Eq. (8.5)) and m now the order parameter. Below a second-order transition, the order parameter follows a critical power law. As discussed above, it is characterized by a power law with a critical exponent, $m = m_0 \tau_r^\beta$, with the reduced temperature $\tau_r \equiv (T_c - T)/T_c$ and $\beta > 0$. It follows that the Seebeck coefficient should also include a critical exponent contribution and be of the form

$$\alpha = \alpha_0 + \alpha_1 \tau_r^r. \tag{8.25}$$

In which α_0 and α_1 may have a separate temperature dependence.

Laguesse *et al.*[50] suggested that Seebeck should have a critical exponent of $r = 1 - \lambda$, where λ is the critical exponent for heat capacity. As λ is less than unity, Laguesse *et al.* predicted r to be greater than 0. This would imply no Seebeck peak and thereby contradict both our observations in Cu_2Se[12] and also the critical exponent of $r = -1$ that Laguesse *et al.* measured in $YBa_2Ca_3O_{7-y}$.[50]

If h is large (relative to τ_r^Δ, where Δ is the gap exponent), then the critical exponent for Seebeck will be larger than $1 - \lambda$. The free energy can be expressed as.[22]

$$f_{\text{2nd}} = \begin{cases} f_0 \tau_r^{2-\lambda} & h \text{ small}, \\ f_1 \tau_r^\beta & h \text{ large}. \end{cases} \tag{8.26}$$

From the above expression, a form of $\frac{\partial f}{\partial n}$ can be obtained:

$$\alpha_{\text{order-entropy}} = -\frac{\partial f_{\text{2nd}}}{q \partial n} = \begin{cases} \alpha_1 \tau_r^{1-\lambda} \dfrac{\partial T_c}{\partial n} & h \text{ small}, \\[2mm] \alpha_2 \tau_r^{\beta-1} \dfrac{\partial T_c}{\partial n} & h \text{ large}. \end{cases} \tag{8.27}$$

From Eq. (8.27), (8.27) has two parts, but they are not labeled 8.27a and 8.27b if h is large enough, the critical exponent for Seebeck is $r = \beta - 1$. As β is typically a small fraction of unity, r should be slightly more than -1. This critical exponent is consistent with that measured in both our work on Cu_2Se and the work by Laguesse *et al.* on $YBa_2Ca_3O_{7-y}$.[50]

The form of Eq. (8.27) can also be determined from our Eq. (8.24) and the definition of $m(T)$:

$$\frac{\partial m}{\partial n} = m_0 \tau_r^{\beta-1} \frac{T}{T_c^2} \frac{\partial T_c}{\partial n}. \tag{8.28}$$

As per our discussion of the first-order transition, the temperature gradient will induce $h \neq 0$. Therefore, $\left(\frac{\partial f}{\partial m}\right)_{T,n} = h$ is nonzero. If this is applied with Eq. (8.28) to Eq. (8.24), the form and critical exponent of Eq. (8.27) are obtained.

8.4.1 Ion-mediated Enhancement

When considering the specific case of super-ionic transitions, ionic transport properties may function as a convenient and metrologically tractable proxy for measurements of the order parameter. A super-ionic transition is a disordering of mobile ions those result in a substantial change in ionic transport properties. The enhancements in ionic transport may be much more significant than those of electronic transport, and the phase transition may act on the electrons indirectly through the ions. In this case, we have

$$\frac{\partial T_c}{\partial n} = \frac{\partial T_c}{\partial n_i} \frac{\partial n_i}{\partial n}, \tag{8.29}$$

where n_i is the concentration of mobile ions. If ionic transport is directly enhanced by the phase transition, $\frac{\partial T_c}{\partial n_i}$ will be nonzero. Such a variation has been observed in both Cu_2Se[25] and other super-ionics.[51,52] Electrons and ions may interact through both chemical and electrostatic processes; their co-transport interaction would indicate a significant value for $\frac{\partial n_i}{\partial n}$. Polarization measurements of some oxygen conductors have revealed such coupled transport,[53,54] which has been explained as being due to a long-range electrostatic interaction changing the effective charge of the transported ions.[55] As $\frac{\partial T_c}{\partial n}$ and $\frac{\partial \alpha}{\partial n}$ may be obtainable via gated transport measurements,[56] future studies may be able to precisely test Eq. (8.27b).

If instead of considering the extra transport as due to the co-transport of the order parameter, we had taken it as due to the co-transport of the ions, then we would have the following entropy production equation:

$$T\dot{S} = -J_s \cdot \nabla T - J_e \cdot \nabla\tilde{\mu}_e - J_i \cdot \tilde{\mu}_i. \tag{8.30}$$

The corresponding Onsager relationships are[5]

$$\begin{bmatrix} J_e \\ J_i \\ J_s \end{bmatrix} = - \begin{bmatrix} ^3L_{ee} & ^3L_{is} & ^3L_{es} \\ ^3L_{ie} & ^3L_{ii} & ^3L_{is} \\ ^3L_{se} & ^3L_{si} & ^3L_{ss} \end{bmatrix} \begin{bmatrix} \nabla\tilde{\mu}_e \\ \nabla\tilde{\mu}_e \\ \nabla T \end{bmatrix}. \tag{8.31}$$

If the Seebeck coefficient is measured in an open circuit system with ion-blocking electrodes (i.e., $J_e = 0$, $J_i = 0$), then

$$\alpha = \frac{\nabla V}{\nabla T}_{J_e=0, J_i=0} = \frac{-1}{q} \frac{^3L_{se}}{^3L_{ee}} \left(\frac{1 - \frac{^3L_{iq}}{^3L_{ii}} \frac{^3L_{ie}}{^3L_{eq}}}{1 - \frac{^3L_{ie}}{^3L_{ii}} \frac{^3L_{ie}}{^3L_{ee}}} \right). \tag{8.32}$$

The first term in the numerator is the "band structure" electronic Seebeck (α_e), as it is equivalent to the Seebeck in a system in which the ion conductivity, specified by L_{ii}, was zero. The second term in the numerator is the Seebeck enhancement due to the ions. The term in the denominator is an ionic drag term. The ionic drag term will be small because $L_{ii} \ll L_{ee}$, and L_{ie} is even in the largest measured case only of the order of L_{ii}. If L_{is} and L_{ie} are significant compared to L_{es} and L_{ee}, the analysis of only electronic properties will lead to an incomplete description of the thermoelectric properties.

The first condition for the ion transport enhancing Seebeck is that $\frac{L_{ie}}{L_{ee}} = \frac{\sigma_{ie}}{\sigma_{ee}}$ be large. $\frac{L_{ie}}{L_{ee}} = \frac{\sigma_{ie}}{\sigma_{ee}}$ is the ratio between ions transported and electrons transported under a gradient in $\tilde{\mu}_e$ but no gradient in $\tilde{\mu}_i$. Mixed ion–electron conductors can show an electron–ion transport coupling. Electrons and ions may interact through both chemical and electrostatic processes. Polarization measurements of some oxygen conductors have revealed such a coupling,[53] which has been explained as due to a long-range electrostatic interaction changing the effective charge of the transported ions.[55]

An observation of not only structural entropy change at the phase transition but also of structural entropy transport is given by Korzhuev and Laptev;[57] they measured a sharp peak in the thermodiffusion of Cu^0 in Cu_2Se at the 410 K phase transition. From this, they calculated a heat of transport of $Q^*_{Cu^0}$ of Cu atoms of 1 eV. The conservation of particles and charge requires that

$$Q^*_{Cu^0} = Q^*_{Cu^+} - Q^*_p = qT\left(\alpha_{Cu^+} - \alpha_p\right). \tag{8.33}$$

Therefore, $Q^*_{Cu^0} = 1\,eV$ corresponds to $\alpha_{Cu^+} = 2500\,\mu VK^{-1}$ at the phase transition. If even 2% small fraction of this copper entropy were co-transported with Cu_2Se, then also the Seebeck and zT enhancement observed would be completely explained.

Mechanistically, this may function through a dependence of the occupation of soft modes (e.g., the Zn_4Sb_3 rattler)[46] on the concentration of Ag^+ and Cu^+.

8.5 Conclusion

Critically, increased entropy may be present in other material systems. $AgCrSe_2$, which is also an ion transporting material, shows a small Seebeck enhancement near a phase transition,[58] though there is no evidence that this is associated with critical phenomena. The temperature of the continuous transition in the CuI–AgI system shows a composition dependence.[52] Work these authors published on $Cu_{1.97}Ag_{0.03}Se$ showed an extended and larger zT peak than that observed in Cu_2Se,[12] while the iodine substitution of copper was shown to significantly reduce the temperature of the phase transition.[18]

The effect on Seebeck from the order entropy is likely not limited to mixed ion–electron conductors. Any material in which the entropy associated with a phase transition might be coupled to transport is a candidate. For example, the magnetic ordering phase transition associated with giant magneto-resistance is often accompanied by a corresponding significant Seebeck change.[59] Applying a magnetic field to these materials induces ordering and results in a corresponding reduction in Seebeck.

In order to understand and engineer this phenomenon, substantial future work needs to be done. The ionic properties, both the conductivity[57] and the Seebeck,[60] may need to be measured and considered when engineering these materials. Further synchotron and neutron crystallographic work may be able to uncover the structure and order parameter. The Onsager coefficient analysis, as discussed previously, may be used to relate separate measurements of ion and electronic properties, in order to directly test the hypothesis of ion-mediated or perhaps even order-mediated Seebeck enhancement.

The broader concept of couple entropy transport may be helpful for understanding polymer and organic thermoelectric systems. In these systems, electronic transport may be related to vibrational or configurational distortions of the materials structure.[61] Polaronic transport is believed important to the thermoelectrics of organic semiconductors.[62] The three-element Onsager treatment of Sandbakk *et al.* was developed to describe transport in ion-conducting membranes.[48] The thermodynamic description as described earlier may be able to supplement existing first principles and kinetics descriptions of these systems, and more importantly, give phenomenological relationships between transport measurements performed under differing experimental conditions.

The best thermoelectric performing materials in this class of compound are yet to be synthesized. Through such future work, a greater understanding of the excellent thermoelectric properties of the ordered phases of super-ionic materials may be understood and engineered.

8.6 Appendix

8.6.1 Seebeck metrology and stability

To minimize the error and increase the resolution of the Seebeck data near the phase transition, a modification to the standard Seebeck metrology was made. I named this technique the "Ramp Seebeck" technique. While in the Oscillation Seebeck technique, $\bar{T}$ is held nominally constant while ΔT is varied and the resultant voltage measured, in the Ramp Seebeck technique, ΔT is held nominally fixed while the $\bar{T}$ is steadily increased or decreased.

During each ramp, the $(\Delta T, V)$ pairs are measured continuously. In our apparatus, the effective $\bar{T}$ steps between data points were on average 0.25 K when ramping at 15 K min^{-1}. Each separate ramp has a slightly different set of $\bar{T}$ values. The $(\bar{T}, \Delta T, V)$ data sets from each ramp were interpolated onto the same $\bar{T}$ values for comparison. The spacing of the new $\bar{T}$ was the average temperature step between data points, as that sets a resolution limit on the measurement procedure.

This process is illustrated in Fig. 8.14, in which the coefficient is plotted point by point without compensation for the voltage offset. If data from a single $\bar{T}$ are explicitly plotted, then a raw Seebeck plot (V versus ΔT) is created, from which a single $(\bar{T}, \alpha)$ point may be determined. This process is performed at all temperatures using the combined heating and cooling data, and is illustrated in Fig. 8.14.

The proof of the superiority of the multiramp method for this problem is its superior results. In Fig. 8.15, we have shown the Seebeck coefficient and dark temperature measured on Cu_2Se by both

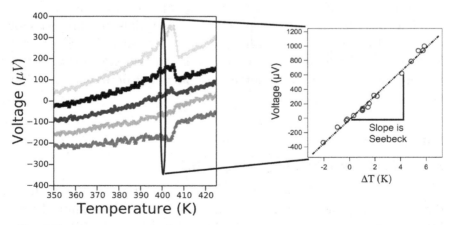

Figure 8.14. Data from multiple ramp sequences are combined point by point to create ΔV versus ΔT from which point by point Seebeck values may be extracted. Adapted from supplemental materials presented in Ref. [12].

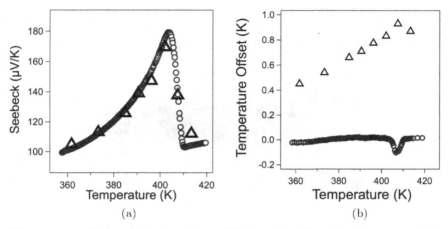

Figure 8.15. Comparison of oscillation (black triangles) and multiramp method data (blue circles) for Cu$_2$Se in proximity to its phase transition. The ramp data are superior. Adapted from supplemental materials presented in Ref. [12].

the oscillation and the ramp methods. While the oscillation method shows the general trend of the anomalous Seebeck peak, it lacks sufficient resolution and clarity to fully describe the phase transition region. The temperature intercept is also of much smaller magnitude during the ramp measurement. The reason for this felicitous situation is unclear, but it is possible that the multiple ramps through the phase transition temperature allow the thermal contacts to stabilize into a better position.

The Seebeck coefficient of Cu$_2$Se just below its phase transition temperature is very stable, indicating that it is a steady-state property of the material; see Figure 8.16. The sample was held at an average temperature of 390 K and a temperature difference of 16 K for 13 h. The measured thermopower, 152 μV K^{-1}, varied by less than 1% during this time period. The predicted average thermopower for this range was calculated by the integration of the data shown in Fig. 8(a) as 143 μV K^{-1}. This 6% discrepancy between the predicted and measured values is consistent with the repetition error of Seebeck measurements. As the discrepancy would suggest that Seebeck and zT are underestimated, it does not undermine the conclusion of phase-transition-enhanced thermopower.

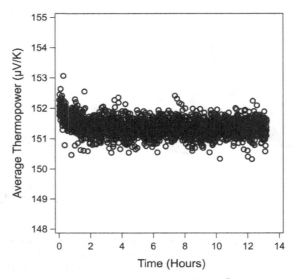

Figure 8.16. Measured Seebeck voltage of Cu$_2$Se at $\bar{T} = 390$ k and $\Delta T = 16$ K. There was a variation of less than 1% in the measured values. After the temperature gradient stabilized there was no variation. Adapted from supplemental materials presented in Ref. [12].

8.6.2 Degradation testing of Cu$_2$Se

A major question for the future development of ionic conducting thermoelectrics is whether these materials will be stable under device conditions of sustained temperature gradients and currents. Both copper and silver show fast interdiffusion in many materials.[63] Therefore, there is strong concern about the decomposition of these materials when operated under the device condition of steady-state current operation. Indeed, Cu$_{1.97}$Ag$_{0.03}$Se and Cu$_2$Se are empirically known to be unstable under the current density of 10 A cm^{-2} at elevated temperatures in their super-ionic phase.[64] However, while these issues may occur at elevated temperatures, my empirical testing shows no evidence of electrochemical degradation at the phase transition temperature.

In order to understand these results and test their applicability to stoichiometric Cu$_2$Se, we performed a short electro-migration experiment. A piece of copper selenide with a cross-sectional area of

$11.47\,\mathrm{mm}^2$ and length of $10.7\,\mathrm{mm}$ was placed in a modified Seebeck apparatus designed to also apply the current. This Seebeck apparatus allows for the simultaneous measurement of voltage and temperature and the application of a current of up to 10 A. The sample was placed under a thermal gradient of $275\,\mathrm{K}$ for $16\,\mathrm{h}$ with $T_h = 795\,\mathrm{K}$ and $T_c = 520\,\mathrm{K}$. The sample was not visibly affected. The current was varied from $0\,\mathrm{A}$ to $10\,\mathrm{A}$ under the same thermal gradient as before. The resistivity did not depend upon the applied current, in contradiction to the General Atomics reports.[65] The sample was then run under the same temperature gradient and in current continuously for $24\,\mathrm{h}$. The magnitude of this current was such that the voltage it induced was half the thermoelectric voltage. The direction was such that it transported Cu^+ in the opposite direction of thermal diffusion; Cu^+ should electromigrate to the hot side and thermally diffuse to the cold side. The applied current was 1 A. The applied current density was $j = 9\,\mathrm{A\,cm^{-2}}$. There was no noticeable degradation of the average Seebeck coefficient over this period; however, when the sample was removed, visual inspection showed that it had undergone deformation at the hot end. Copper residue was visible on the face that had been against the hot-side heater and the current sink.

Fortunately, these problems are eliminated below the phase transition temperature. The author performed high current density plate-out tests at three additional temperatures, the results of which are summarized in Table 8A.1. Currents and durations far in excess of that applied above the phase transition resulted in no observed copper plate out or altering of the transport properties. In general, the super-ionic phase, in addition, of course, to its elevated

Table 8A.1. Electromigration experiments on Cu_2Se (no electomigrative loss of copper was observed below the phase transition temperature).

$\bar{T}_{(K)}$	$\Delta T_{(K)}$	J (A cm^{-2})	Duration (h)	Plate out?
658	275	9	16	Yes
404	23	21.7	260	No
387	19	29.3	70	No

ionic conductivity, is thermophysically quite distinct from the low-temperature phase. It shows ultrafast chemical diffusion[34,57,66] and superplasticity.[67] In this context, it is unsurprising that the ion-ordered phases do not show the same thermophysical instability as the super-ionic phase.

References

1. G. J. Snyder and E. S. Toberer, Complex thermoelectric materials, *Nat. Mater.*, **7**(2), 105–114 (2008).
2. G. D. Mahan and J. O. Sofo, The best thermoelectric, *Proc. Natl. Acad. Sci. USA*, **93**(15), 7436–7439 (1996).
3. Y. Pei, H. Wang, and G. J. Snyder, Band engineering of thermoelectric materials, *Adv. Mater.*, **24**(46), 6125–6135 (2012).
4. L. Onsager, Reciprocal relations in irreversible processes. I, *Phys. Rev.*, **37**(4), 405–426 (1931); H. B. Callen, The application of Onsager's reciprocal relations to thermoelectric, thermomagnetic, and galvanomagnetic effects, *Phys. Rev.* **73**(11), 1349–1358 (1948); C. A. Domenicali, Irreversible thermodynamics of thermoelectricity, *Rev. Mod. Phys.* **26**(2), 237 (1954).
5. S. R. DeGroot and P. Mazur, *Non-Equillibrium Thermodynamics.* Dover Publications: New York, 1962.
6. C. Goupil, W. Seifert, K. Zabrocki, E. Muller, and G. J. Snyder, Thermodynamics of thermoelectric phenomena and applications, *Entropy*, **13**(8), 1481–1517 (2011).
7. D. Emin, Enhanced Seebeck coefficient from carrier-induced vibrational softening, *Phys. Rev. B*, **59**(9), 6205–6210 (1999); T. L. Aselage, D. Emin, S. S. McCready, and R. V. Duncan, Large enhancement of boron carbides' Seebeck coefficients through vibrational softening, *Phys. Rev. Lett.* **81**(11), 2316 (1998).
8. W. Koshibae, K. Tsutsui, and S. Maekawa, Thermopower in cobalt oxides, *Phys. Rev. B*, **62**(11), 6869–6872 (2000).
9. W. Koshibae and S. Maekawa, Effects of spin and orbital degeneracy on the thermopower of strongly correlated systems, *Phys. Rev. Lett.*, **87**(23) (2001).
10. J. L. Opsal, B. J. Thaler, and J. Bass, Electron–phonon mass enhancement in thermoelectricity, *Phys. Rev. Lett.*, **36**(20), 1211 (1976).
11. Y. Y. Wang, N. S. Rogado, R. J. Cava, and N. P. Ong, Spin entropy as the likely source of enhanced thermopower in $Na_xCo_2O_4$, *Nature* **423**(6938), 425–428 (2003).

12. D. R. Brown, T. Day, K. A. Borup, S. Christensen, B. B. Iversen, and G. J. Snyder, Phase transition enhanced thermoelectric figure-of-merit in copper chalcogenides, *APL Mater.*, **1**(5), 052107 (2013).
13. C. N. R. Rao and K. J. Rao, *Phase Transitions in Solids: An Approach to the Study of the Chemistry and Physics of Solids*, McGraw-Hill: New York, 1978.
14. J. B. Boyce and B. A. Huberman, Superionic conductors — transitions, structures, dynamics, *Phys. Rep. -Rev. Sec. Phys. Lett.*, **51**(4), 189–265 (1979).
15. S. A. Danilkin, M. Avdeev, T. Sakuma, R. Macquart, and C. D. Ling, Neutron diffraction study of diffuse scattering in Cu_2-delta Se superionic compounds, *J. Alloys Compounds*, **509**(18), 5460–5465 (2011); A. Tonejc and A. M. Tonejc, X-ray-diffraction study on alpha-reversible-beta-phase transition of Cu_2Se, *J. Solid State Chem.*, **39**(2), 259–261 (1981).
16. Z. Vucic, V. Horvatic, and O. Milat, Dilatometric study of nonstoichiometric copper selenide $Cu_{2-x}Se$, *Solid State Ionics*, **13**(2), 127–133 (1984).
17. H. Liu, X. Shi, M. Kirkham, H. Wang, Q. Li, C. Uher, W. Zhang, and L. Chen, Structure-transformation-induced abnormal thermoelectric properties in semiconductor copper selenide, *Mater. Lett.* **93**, 121–124 (2013).
18. H. Liu, X. Yuan, P. Lu, X. Shi, F. Xu, Y. He, Y. Tang, S. Bai, W. Zhang, L. Chen, Y. Lin, L. Shi, H. Lin, X. Gao, X. Zhang, H. Chi, and C. Uher, Ultrahigh thermoelectric performance by electron and phonon critical scattering in $Cu_2Se_{1-x}I_x$, *Adv. Mater.*, **25**(45), 6607–6612 (2013).
19. P. Ehrenfest and T. Ehrenfest, *The Conceptual Foundations of the Statistical Approach in Mechanics*, Courier Dover Publications, USA (2002).
20. R. A. Ferrell, N. Menyhard, H. Schmidt, F. Schwabl, and P. Szepfalusy, Fluctuations and lambda phase transition in liquid helium, *Ann. Phys.*, **47**(3), 565–613 (1968).
21. A. B. Pippard, XLVIII. Thermodynamic relations applicable near a lambda-transition, *Philos. Mag.*, **1**(5), 473–476 (1956).
22. M. Kardar, *Statistical Physics of Fields*, Cambridge University Press, Cambridge, UK, 2007.
23. W. J. Pardee and G. D. Mahan, Disorder and ionic polarons in solid electrolytes, *J. Solid State Chem.*, **15**(4), 310–324 (1975); N. Hainovsky and J. Maier, Simple phenomenological approach to premelting and sublattice melting in Frenkel disordered ionic crystals, *Phys. Rev. B*, **51**(22), 15789–15797 (1995).
24. R. D. Heyding, The copper/selenium system, *Canadian J. Chem.*, **44**(10), 1233–1236 (1966).

25. Z. Vučić, O. Milat, V. Horvatić, and Z. Ogorelec, Composition-induced phase-transition splitting in cuprous selenide, *Phys. Rev. B*, **24**(9), 5398–5401 (1981).

26. P. Rahlfs, The cubic high temperature modificators of sulfides, selenides and tellurides of silver and of uni-valent copper, *Zeitschrift für physikalische Chemie. Abteilung B, Chemie der Elementarprozesse, Aufbau der Materie*, **31**(3), 157–194 (1936).

27. K. Yamamoto and S. Kashida, X-ray study of the cation distribution in Cu_2Se, $Cu_{1.8}Se$ and $Cu_{1.8}S$ — analysis by the maximum-entropy method, *Solid State Ionics*, **48**(3–4), 241–248 (1991).

28. S. Danilkin, Diffuse scattering and lattice dynamics of superionic copper chalcogenides, *Solid State Ionics*, 180(6–8), 483–487 (2009); S. A. Danilkin, M. Avdeev, M. Sale, and T. Sakuma, Neutron scattering study of ionic diffusion in Cu–Se superionic compounds, *Solid State Ionics*, **225**, 190–193 (2012).

29. S. Kashida and J. Akai, X-ray-diffraction and electron-microscopy studies of the room-temperature structure of Cu_2Se, *J. Phys. C — Solid State Phys.*, **21**(31), 5329–5336 (1988).

30. O. Milat, Z. Vucic, and B. Ruscic, Superstructural ordering in low-temperature phase of superionic Cu_2Se, *Solid State Ionics*, **23**(1–2), 37–47 (1987).

31. B. Vengalis, K. Valacka, N. Shiktorov, and V. Yasutis, The state of Cu_2-delta-Se below the superionic phase-transition temperature, *Fizika Tverdogo Tela*, **28**(9), 2675–2679 (1986).

32. A. B. Thompson and E. H. Perkins, Lambda transitions in minerals. In *Thermodynamics of Minerals and Melts*, Springer: Berlin, 1981, pp. 35–62.

33. U. T. Hochli and J. F. Scott, Displacement parameter, soft-mode frequency, and fluctuations in quartz below its alpha-beta phase transition, *Phys. Rev. Lett.*, **26**(26), 1627–& (1971).

34. M. A. Korzhuev, N. K. Abrikosov, and I. V. Kuznetsova, Isolation of moving copper from the $Cu_{2-x}Se$ under the pressure effect, *Pisma V Zhurnal Tekhnicheskoi Fiziki*, **13**(1), 9–14 (1987).

35. P. Navard and J. M. Haudin, The height of DSC phase transition peaks, *J. Thermal Anal. Calorimetry*, **29**(3), 405–414 (1984).

36. W. Weppner, C. Lichuan, and W. Piekarczyk, Electrochemical determination of phase-diagrams and thermodynamic data of multicomponent systems, *Zeitschrift Fur Naturforschung Section A-A J. Phys. Sci.*, **35**(4), 381–388 (1980).

37. M. A. Korzhuev, The entropy of fusion and the growth-morphology of superionic $Cu_{2-x}Se$ crystals, *Physica Status Solidi A – Appl. Res.*, **127**(1), K1–K3 (1991).

38. M. Horvatic and Z. Vucic, Dc ionic-conductivity measurements on the mixed conductor $Cu_{2-x}Se$, *Solid State Ionics*, **13**(2), 117–125 (1984).

39. Z. Vucic, V. Horvatic, and Z. Ogorelec, Influence of the cation disordering on the electronic conductivity of superionic copper selenide, *J. Physics C — Solid State Physics*, **15**(16), 3539–3546 (1982).

40. F. El Akkad, B. Mansour, and T. Hendeya, Electrical and thermoelectric properties of Cu_2Se and Cu_2S, *Mater. Res. Bull.*, **16**(5), 535–539 (1981); H. Liu, X. Shi, F. Xu, L. Zhang, W. Zhang, L. Chen, Q. Li, C. Uher, T. Day, and G. J. Snyder, Copper ion liquid-like thermoelectrics, *Nat. Mater.*, **11**(5), 422–425 (2012); Z. Ogorelec, and B. Celustka, Thermoelectric power and phase transitions of nonstoichiometric cuprous selenide, *J. Phys. Chem. Solids*, **27**(3), 615–617 (1966); K. Okamoto, Thermoelectric power and phase transition of Cu_2Se, *Jpn. J. Appl. Phys.*, **10**(4), 508 (1971); X.-X. Xiao, W.-J. Xie, X. -F. Tang, and Q.-J. Zhang, Phase transition and high temperature thermoelectric properties of copper selenide $Cu_{2-x}Se$ ($0 \leq x \leq 0.25$), *Chin. Phys. B*, **20**(8), 087201 (2011); G. Bush and P. Junod, Relations between the crystal structure and electronic properties of the compounds Ag 2S, Ag 2Se, Cu 2Se, *Helvetica Physica Acta*, **32**(6–7), 567–600 (1959).

41. Z. Vucic and Z. Ogorelec, Unusual behavior of nonstoichiometric cuprous selenide at the transition to the superionic phase, *Philos. Mag. B — Phys. Cond. Matter Stat. Mech. Electron. Opt. Magn. Prop.*, **42**(2), 287–296 (1980).

42. M. A. Korzhuev, V. F. Bankina, I. G. Korolkova, G. B. Sheina, and E. A. Obraztsova, Doping effects on mechanical-properties and micro-hardness of superionic copper selenide $Cu_{2-x}Se$, *Physica Status Solidi A - Appl. Res.*, **123**(1), 131–137 (1991).

43. F. L. Lederman, M. B. Salamon, and H. Peisl, Evidence for an order-disorder transformation in solid electrolyte Rb Ag-4i-5, *Solid State Commun.*, **19**(2), 147–150 (1976).

44. G. K. H. Madsen and D. J. Singh, BoltzTraP. A code for calculating band-structure dependent quantities, *Comp. Phys. Commun.* **175**(1), 67–71 (2006).

45. D. M. Rowe, *Materials, Preparation, and Characterization in Thermoelectrics*. Taylor & Francis, New York, 2012.

46. W. Schweika, R. P. Hermann, M. Prager, J. Persson, and V. Keppens, Dumbbell rattling in thermoelectric zinc antimony, *Phys. Rev. Lett.*, **99**(12), 125501 (2007).

47. B. M. Askerov, *Electron transport phenomena in semiconductors*. (World Scientific, 1994), pp. 168–255.

48. K. D. Sandbakk, A. Bentien, and S. Kjelstrup, Thermoelectric effects in ion conducting membranes and perspectives for thermoelectric energy conversion, *J. Membr. Sci.*, **434**(0), 10–17 (2013).

49. H. Wang, W. D. Porter, H. Böttner, J. König, L. Chen, S. Bai, T. M. Tritt, A. Mayolet, J. Senawiratne, and C. Smith, Transport properties of bulk thermoelectrics — an international round-robin study, part I: Seebeck coefficient and electrical resistivity, *J. Elect. Mater.*, **42**(4), 654–664 (2013).

50. M. Laguesse, A. Rulmont, Ch Laurent, S. K. Patapis, H. W. Vanderschueren, P. Tarte, and M. Ausloos, Thermoelectric power and magneto Seebeck-effect near the critical temperature of granular ceramic oxide superconductors $Y1Ba_2Ca_3O_{7-y}$, *Solid State Commun.*, **66**(4), 445–450 (1988).

51. Y. Kowada, Y. Yamada, M. Tatsumisago, T. Minami, and H. Adachi, Variation of electronic state of AgI-based superionic conductors with movement of Ag ions, *Solid State Ionics*, **136**, 393–397 (2000).

52. M. Kusakabe, Y. Ito, and S. Tamaki, The specific heat of the solid electrolyte system CuI-AgI, *J. Phys. -Condensed Matter*, **8**(37), 6851–6856 (1996).

53. H.-I. Yoo, J.-H. Lee, M. Martin, J. Janek, and H. Schmalzried, Experimental evidence of the interference between ionic and electronic flows in an oxide with prevailing electronic conduction, *Solid State Ionics*, **67**(3–4), 317–322 (1994); J. H. Lee and H. I. Yoo, Electrochemical study of the cross effect between ion and electron flows in semiconducting CoO, *J. Electrochem. Soc.*, **141**(10), 2789–2794 (1994).

54. C. Chatzichristodoulou, W.-S. Park, H.-S. Kim, P. V. Hendriksen, and H.-I. Yoo, Experimental determination of the Onsager coefficients of transport for $Ce_{0.8}Pr_{0.2}O_2$-[small delta], *Phys. Chem. Chem. Phys.* **12**(33), 9637–9649 (2010).

55. H. S. Kim and H. I. Yoo, Compilation of all the isothermal mass/charge transport properties of the mixed conducting La_2NiO_4+delta at elevated temperatures, *Phys. Chem. Chem. Phys.*, **13**(10), 4651–4658 (2011).

56. V. Sandomirsky, A. V. Butenko, R. Levin, and Y. Schlesinger, Electric-field-effect thermoelectrics, *J. Appl. Phys.*, **90**(5), 2370–2379 (2001).

57. M. A. Korzhuev and A. V. Laptev, Thermodiffusion and piezodiffusion effects in superionic copper selenide, *Fizika Tverdogo Tela*, **29**(9), 2646–2650 (1987).

58. F. Gascoin and A. Maignan, Order–disorder transition in AgCrSe(2): A new route to efficient thermoelectrics, *Chem. Mater.*, **23**(10), 2510–2513 (2011).

59. M. Jaime, M. B. Salamon, K. Pettit, M. Rubinstein, R. E. Treece, J. S. Horwitz, and D. B. Chrisey, Magnetothermopower in La$_{0.67}$Ca$_{0.33}$MnO$_3$ thin films, *Appl. Phys. Lett.*, **68**(11), 1576–1578 (1996); C. J. Liu, C. S. Sheu, T. W. Wu, L. C. Huang, F. H. Hsu, H. D. Yang, G. V. M. Williams, and C. J. C. Liu, Magnetothermopower and magnetoresistivity of RuSr(2)Gd(1−x)La(x)Cu(2)O(8) (x = 0,0.1), *Phys. Rev. B*, **71**(1), 14502 (2005).

60. M. Kh Balapanov, I. B. Zinnurov, and G. R. Akmanova, The ionic Seebeck effect and heat of cation transfer in Cu$_{2−\delta}$Se superionic conductors, *Phys. Solid State*, **48**(10), 1868–1871 (2006).

61. J. L. Brédas, J. P. Calbert, D. A. da Silva Filho, and J. Cornil, Organic semiconductors: A theoretical characterization of the basic parameters governing charge transport, *Proc. Natl Acad. Sci.* **99**(9), 5804–5809 (2002).

62. K. P. Pernstich, B. Rössner, and B. Batlogg, Field-effect-modulated Seebeck coefficient in organic semiconductors, *Nat. Mater.*, **7**(4), 321–325 (2008).

63. B. F. Dyson, T. Anthony, and D. Turnbull, Interstitial diffusion of copper and silver in lead, *J. Appl. Phys.*, **37**(6), 2370–2374 (1966); Andrei A Istratov, Christoph Flink, Henry Hieslmair, Eicke R. Weber, and Thomas Heiser, Intrinsic diffusion coefficient of interstitial copper in silicon, *Phys. Rev. Lett.*, **81**(6), 1243 (1998).

64. D. R. Brown, T. Day, T. Caillat, and G. J. Snyder, Chemical stability of (Ag,Cu)2Se: A historical overview, *J. Electron. Mater.*, **42**(7), 2014–2019 (2013); G. Dennler, R. Chmielowski, S. Jacob, F. Capet, P. Roussel, S. Zastrow, K. Nielsch, I. Opahle, and G. K. H. Madsen, Are binary copper sulfides/selenides really new and promising thermoelectric materials?, *Adv. Ener. Mater.*, n/a-n/a (2014).

65. N. B. Elsner, J. Chin, G. H. Reynolds, J. H. Norman, J. C. Bass, and H. G. Staley, Report No. GA-A-16584; Other: ON: DE82015077 United States10.2172/5437033Other: ON: DE82015077Thu Jul 05 07:24:33 EDT 2012NTIS, PC A07/MF A01.GA; INS-82-013511; NTS-82-011522; ERA-07-047144; EDB-82-114375English, 1981.

66. M. A. Korzhuev, Dufour effect in superionic copper selenide, *Phys. Solid State*, **40**(2), 217–219 (1998).

67. N. N. Sirota, M. A. Korzhuev, M. A. Lobzov, N. K. Abrikosov, and V. F. Bankina, Superplasticity of beta-phase of copper selenide possessing the superionic conduction, *Doklady Akademii Nauk SSSR*, **281**(1), 75–77 (1985).

Index

Printed in the United States
By Bookmasters